Fuzzy
Switching
and
Automata:
Theory and
Applications

Computer Systems Engineering Series

DOUGLAS LEWIN, *Editor*

Fuzzy Switching and Automata:

Theory and Applications

Abraham Kandel
Samuel C. Lee

Crane Russak · *New York*
Edward Arnold · *London*

Fuzzy Switching and Automata
Published in the United States by
Crane, Russak & Company, Inc.
347 Madison Avenue
New York, New York 10017
ISBN 0-8448-1020-7
LC 76-29148

Published in Great Britain by
Edward Arnold (Publishers) Ltd.
41 Bedford Square
London WC1B 3DP
ISBN 0 7131 2670 1

Printed in the United States of America

Contents

To our children

Sharon, Gill, Adi, Vivian, and Jennifer

Preface

During the past decade, there has been a considerable growth of interest and research results in fuzzy theory in general, and in fuzzy switching and automata theory in particular. A large volume of papers and reports on fuzzy theory is published each year in journals and conference proceedings in the United States and abroad. There is need for a book on fuzzy switching and automata theory to help those who would like to learn the theory in a relatively short time, or improve the knowledge they already have. This book is an attempt to meet such a need.

Many people recognize that today's computer scientists and system engineers must be familiar with fuzzy theory and its applications, since the theory is a mathematical tool that can be applied to solving practical real-world soft-science problems in many scientific fields—from pattern recognition to neural networks, from language theory to logic design. This book includes many of our recent research results in the field. The presentation is kept concise in order to make room for the coverage of major applications and also to keep the book to reasonable length.

A brief description of the book by chapter follows.

Chapter 1 presents a review of set theory and of two important algebraic systems—lattices and Boolean algebra—which are the foundations of switching and automata theory.

Chapter 2 is a collection of basic results of fuzzy set theory, algebra and logic, intended for the reader who finds this material completely or partially novel.

Chapter 3 is concerned with the study of fuzzy switching functions, their minimization, decomposition, and other related topics.

Automata have been used as models for sequential circuits and discrete systems including both engineering and biological systems. Formal languages have been used to describe natural and computer languages, as well as in the analysis of pictorial and speech patterns. Extensions to fuzzy automata and fuzzy language have increased the power of both to model systems and languages under imprecision. These aspects of fuzzy theory are described in Chapter 4.

Chapter 5 discusses various applications of the previous concepts. These applications include the investigation of fuzzy neural networks, approximation via fuzzy functions, transient analysis of binary switching systems, and applications to role theory and pattern recognition and classification.

The book is intended for researchers in the field of fuzzy theory and related areas. It can be used as a text for one-term courses in advanced switching and automata theory, and in fuzzy logic and its applications. It assumes that the reader has had an elementary knowledge of switching and automata theory and some degree of mathematical sophistication.

Some of the material contained in this book was used in courses at the New Mexico Institute of Mining and Technology and the Florida State University. We are indebted to our students and peers for criticisms and suggestions.

We are also indebted to all those who contributed to the production of this book. We would like to acknowledge our particular indebtedness to C. L. Chang, M. A. Davis, J. S. Hughes, E. T. Lee, P. N. Marinos, F. P. Preparata, G. W. Schwede, M. G. Thomason, R. T. Yeh, L. Yelowitz, and L. A. Zadeh. Portions of the material in this book are derived from their early investigations in this field.

A bibliography, listing the work of these and many other researchers in the area of fuzzy sets and their applications and related topics is included at the end of the volume.

Dr. Araham Kandel
Associate Professor and Director
of Computer Science
The Florida State University
Tallahassee, FL. 32306

and

Dr. Samuel C. Lee
Professor of Electrical Engineering
University of Oklahoma
Norman, OK.

November 1978

Fuzzy
Switching
and
Automata:
Theory and
Applications

CHAPTER 1

Sets, Lattices, and Boolean Algebras

1.1. INTRODUCTION

One of the most important tools in modern mathematics is the theory of sets. The notation, terminology, and concept of set theory are helpful in studying any branch of mathematics. Every branch of mathematics can be considered as a study of sets of objects of one kind or another. For example; geometry is a study of sets of points. Algebra is concerned with sets of numbers and operations on those sets. Analysis deals mainly with sets of functions. The study of sets and their use in the foundations of mathematics was begun in the latter part of the nineteenth century by the German mathematician George Cantor (1845–1918). Since then set theory has unified many seemingly disconnected ideas. It has helped to reduce many mathematical concepts to their logical foundations in an elegant and systematic way. Moreover, it has not only influenced and enriched almost every branch of mathematics, but has also helped to clarify the relationship between mathematics and philosophy.

A collection of objects, in which nothing special is assumed about the nature of the individual objects, is called a *set*. The individual objects in the collection are called *elements* or *members* of the set, and are said *to belong to* (or *to be contained in*) the set. A group of men, a bunch of flowers, and a sequence of numbers are examples of sets. Here, men, flowers, and numbers are elements or members of these sets. It is important to realise that a set may itself be an element of some other set. For example, a line is a set of points; the set of all lines in the plane is a set of sets of points. In fact a set can be a set of sets of sets and so on. The theory dealing with the (abstract) sets defined in the

1

above manner is called (*abstract* or *conventional*) *set theory,* in contrast to *fuzzy set theory* which will be introduced in Chapter 2. This chapter begins with a review of set theory which includes the introduction of several important classes of sets and their properties.

Two important algebraic systems, lattice and Boolean algebra, are then studied. Both systems are derived from ordered sets. There are two types of lattice, modular and distributive. The complemented distributive lattice, which is a subclass of the distributive lattice, is called Boolean algebra. Every Boolean algebra is a distributive lattice, and every distributive lattice is a modular lattice. In other words, for a lattice to be a Boolean algebra it is necessary that it be distributive. This, in turn, requires that it be modular.

1.2. OPERATIONS ON SETS

Sets are usually denoted by capital letters $A, B, C, \ldots, X, Y, Z$. Elements will be denoted by lower-case letters $a, b, c, \ldots, x, y, z$. The phrase "is an element of" will be denoted by the symbol $\in$. Thus we write $x \in A$ for "x is an element of A". Here, and in analogous situations, we indicate the denial of a statement by use of $|$. Thus, we write $x \notin A$ for "x is *not* an element of A".

There are five ways used to describe a set.

(1) Describe a set by describing the properties of the members of the set.

(2) Describe a set by listing its elements.

(3) Describe a set A by its characteristic function, defined as

$$\mu_A(x) = 1 \quad \text{if} \quad x \in A$$

$$\mu_A(x) = 0 \quad \text{if} \quad x \notin A$$
for all x in A.

(4) Describe a set by a recursive formula. This is to give one element of the set and a rule by which the rest of the elements of the set may be found.

(5) Describe a set by an operation (such as union, intersection, complement, etc.) on some sets.

Example 1.2.1. Describe the set containing all the non-negative integers less than or equal to 5.

Let A denote the set. Set A can be described in the following ways.

(1) $A = \{x \mid x$ is a non-negative integer less than or equal to $5\}$

(2) $A = \{0, 1, 2, 3, 4, 5\}$

(3) $\mu_A(x) = \begin{cases} 1 & \text{for } x = 0, 1, \ldots, 5 \\ 0 & \text{otherwise} \end{cases}$

(4) $A = \{x_{i+1} = x_i + 1, i = 0, 1, \ldots, 4, \text{ where } x_0 = 0\}$

(5) This part is left to the reader as an exercise.

Note that, for a given set, not all the five ways of describing it are always possible. For example, the set of real numbers between 0 and 1 cannot be described by either listing all its elements or by a recursive formula.

In this section, we shall introduce the fundamental operations on sets and the relations among these operations. We begin with the following definitions.

Definition 1.2.1. Let A and B be two sets. A is said to be a *subset* of B if every element of A is an element of B. A is said to be a *proper subset* of B if A is a subset of B and there is at least one element of B which is not in A.

If A is a subset of B, we say A is *contained* in B. Symbolically, we write $A \subseteq B$. If A is a proper subset of B, then we say A is *strictly contained* in B, denoted by $A \subset B$. The containment of sets has the following properties. Let A, B, and C be sets.

(1) $A \subseteq A$

(2) If $A \subseteq B$ and $B \subseteq C$, then $A \subseteq C$

(3) If $A \subseteq B$ and $B \subset C$, then $A \subset C$

(4) If $A \subseteq B$ and $A \not\subseteq C$, then $B \not\subseteq C$

The statement $A \subseteq B$ does not rule out the possibility that $B \subseteq A$. In fact, we have both $A \subseteq B$ and $B \subseteq A$ if, and only if (abbreviated iff), A and B have the same elements. Thus we define the following.

Definition 1.2.2. Two sets A and B are *equal* iff $A \subseteq B$ and $B \subseteq A$. We write $A = B$.

A set containing no elements is called the *empty* set or *null* set, denoted by $\varnothing$. For example, the set of all positive numbers x satisfying the equation $x + 1 = 0$ is an empty set since there are no positive numbers which can satisfy this equation. The empty set is a subset of every set. In other words, $\emptyset \subseteq A$ for every A. This is because there are no elements in $\emptyset$; therefore, every element in $\emptyset$ belongs to A. It is important to note that the sets $\emptyset$ and $\{\emptyset\}$ are very different sets. The former has no elements, whereas the latter has the unique element $\emptyset$. A set containing a single element is called a *singleton.*

Another special set is the *universal* set U. The universal set is defined as the set which contains every other set. Thus, any set is a subset of the universal set. In many discussions some particular set U will be especially important and many subsets of U will be considered. In such a context U may be called the universal set for the discussion. In this case the set U does not contain everything, but rather is just the universal of discourse at a particular time.

We shall now describe three operations on sets, namely, complement, union, and intersection. These operations allow us to construct new sets from given sets. We shall also study the relationships among these operations.

Definition 1.2.3. Let U be the universal set and let A be any set. The *absolute complement* of A, $\overline{A}$, is defined as $\{x \,|\, x \notin A\}$ or, $\{x \,|\, x \in U$ and $x \notin A\}$. If A and B are sets, the *relative complement* of A with respect to B is as shown below.

$$B - A = \{x \,|\, x \in B \text{ and } x \notin A\}$$

It is clear that $\overline{\emptyset} = U$, $\overline{U} = \emptyset$, and that the complement of the complement of A is equal to A.

Definition 1.2.4. Let A and B be two sets. The *union* of A and B is $A \cup B = \{x \,|\, x \in A \text{ or } x \in B \text{ or both}\}$. More generally, if $A_1, A_2, \ldots,$ A_n are sets, then their union is the set of all objects which belong to *at least one* of them, and is denoted by $A_1 \cup A_2 \cup \ldots \cup A_n$, or by $\bigcup_{k=1}^{n} A_k$.

Definition 1.2.5. The *intersection* of two sets A and B is $A \cap B = \{x \,|\, x \in A \text{ and } x \in B\}$. The intersection of n sets $A_1, A_2, \ldots,$ A_n is the set of all objects which belong to *every one* of them, and is denoted by $A_1 \cap A_2 \cap \ldots \cap A_n$, or $\bigcap_{k=1}^{n} A_k$.

Some basic properties of union and intersection of two sets are as follows.

	Union	Intersection
(1) Idempotent:	$A \cup A = A$	$A \cap A = A$
(2) Commutative:	$A \cup B = B \cup A$	$A \cap B = B \cap A$
(3) Associative:	$A \cup (B \cup C) = (A \cup B) \cup C$	$A \cap (B \cap C) = (A \cap B) \cap C$

It should be noted that, in general:

$$(A \cup B) \cap C \neq A \cup (B \cap C)$$

Definition 1.2.6. The *symmetrical difference of two sets A* and *B* is $A \triangle B = \{x \mid x \in A, \text{ or } x \in B, \text{ but not both}\}$. The symmetrical difference of two sets is also called the *Boolean sum* of the two sets.

Definition 1.2.7. Two sets *A* and *B* are said to be *disjoint* if they do not have a member in common, that is to say, if $A \cap B = \emptyset$. More generally, sets $A_1, A_2, \ldots, A_n$ are disjoint if no element belongs to more than one of $A_1, A_2, \ldots,$ and A_n, or $\bigcap_{k=1}^{n} A_k = \emptyset$.

We can easily show the following theorems from the definitions of union, intersection, and complement.

Theorem 1.2.1. (Distributive Laws). Let *A, B,* and *C* be three sets. Then:

$$C \cap (A \cup B) = (C \cap A) \cup (C \cap B)$$

$$C \cup (A \cap B) = (C \cup A) \cap (C \cup B)$$

Theorem 1.2.2. (DeMorgan's Laws). Let *A* and *B* be two sets. Then:

$$\overline{(A \cup B)} = \overline{A} \cap \overline{B}$$

$$\overline{(A \cap B)} = \overline{A} \cup \overline{B}$$

The proofs of these two theorems are left to the reader as exercises.

Exercises

(1) In a city, there is one and only one barber *b*. He only serves for those who do not cut their own hair. Define $S = \{x \mid x \text{ who does not cut his own hair}\}$. Is *b* in *S*? If not, is *b* not in *S*? If *b* is neither in *S* nor not in *S*, then find the reason why this situation occurs.

(2) Write a description of the set containing all the non-negative integers less than or equal to 5 by using an operation (a) union, (b) intersection, (c) complement operating on two sets.

(3) Prove Theorems 1.2.1 and 1.2.2.

(4) Prove the following identities. Let *A, B, C,* and *D* be sets.
(a) $A \cup \emptyset = A; A \cap \emptyset = \emptyset$
(b) $A \cup U = U; A \cap U = A$

(c) $\emptyset - A = \emptyset$

(d) $A \triangle \emptyset = A$

(e) $A - B = A \cap \overline{B}$

(f) $A - (A - B) = A \cap B$

(g) $C \cap D = A - [(A - C) \cup (A - D)]$, where $A \cap C = \emptyset$ and $A \cap D = \emptyset$

(h) $C \cup D = A - [(A - C) \cap (A - D)]$, where $A \cap C = \emptyset$ and $A \cap D = \emptyset$

(i) $A \triangle B = (A - B) \cup (B - A)$

(j) $A \triangle B = (A \cup B) - (A \cap B)$

(5) Verify the following statements.

(a) $A \subset B$ iff $A \cup B = B$ iff $A \cap B = A$ iff $A - B = \emptyset$

(b) $A \subseteq \emptyset$ iff $A = \emptyset$

(6) Which of the following statements are false? Give an example for each false statement to show that it is false.

(a) The properties of the symmetrical difference of two sets is associative.

(b) If A and B are two sets, then $A \cap B \subseteq A \cup B$

(c) If A and B are two sets, then $A - B = B - A$

(d) If A, B, and C are sets, and $A \neq B$ and $B \neq C$, then $A \neq C$

It is often helpful to use a diagram, called a **Venn diagram** (John Venn 1834–1883), to visualize the various properties of the set operations. The universal set is represented by a large square area. Subsets within this universe are represented by circular areas. A summary of set operations and their Venn diagrams is given in Table 1.1.

1.3. CARTESIAN PRODUCTS, RELATIONS, AND FUNCTIONS

In this section we shall be mainly concerned with sets whose elements are ordered pairs. By ordered pair we mean that each set is specified by two objects in a prescribed order. The ordered pair of a and b, with first co-ordinate a and second co-ordinate b, is the set (a,b). We also define that $(a,b) = (c,d)$ if and only if $a = c$ and $b = d$. We are now in a position to define the Cartesian product of sets A and B.

Table 1.1. Venn Diagram of Set Operations

Set Operation	Symbol	Venn Diagram
Set B is contained in set A	$B \subset A$	
The absolute complement of set A	$\overline{A}$	
The relative complement of set B with respect to set A	$A - B$	
The union of sets A and B	$A \cup B$	
The intersection of sets A and B	$A \cap B$	
The symmetrical difference of sets A and B	$A \triangle B$	

Definition 1.3.1. Let A and B be two sets. The *Cartesian product* of A and B is defined as $A \times B = \{(a,b) \mid a \in A \text{ and } b \in B\}$. More generally, the Cartesian product of n sets $A_1, A_2, \ldots, A_n$ is defined:

$$A_1 \times A_2 \times \ldots \times A_n = \{(a_1, a_2, \ldots, a_n) \mid a_i \in A_i, i = 1, 2, \ldots, n\}$$

The symbol $(a_1, a_2, \ldots, a_n)$ is called an *ordered n-tuple*.

Example 1.3.1. Let $A = \{0,1,2\}$ and $B = \{3,4\}$. Then:

$$A \times B = \{(0,3),(0,4),(1,3),(1,4),(2,3),(2,4)\}$$

$$A \times A = \{(0,0),(0,1),(0,2),(1,0),(1,1),(1,2),(2,0),(2,1),(2,2)\}$$

Example 1.3.2. Let R^1 be the set of real numbers. Then the Cartesian product $R^1 \times R^1 = \{(x,y) \mid x \text{ and } y \text{ are real numbers}\}$.

Exercises

(1) A, B, C, and D are sets. Prove the following identities.
(a) $A \times (B \cup C) = (A \times B) \cup (A \times C)$
(b) $(A \times B) \cap C = (A \times C) \cap (B \times C)$
(c) $(A \cap B) \times (C \cap D) = (A \times C) \cap (B \times D)$
(d) $(A \cup B) \times (C \cup D) = (A \times C) \cup (B \times D)$

(2) Determine which of the following expressions are true.
Which are false? Give your reasons.
(a) $A \times B = B \times A$. If false, under what circumstances will it be true?
(b) $(A - B) \times C = (\underline{A} \times C) - (\underline{B} \times C)$
(c) $\overline{(A \cap B)} \times C = (\overline{A} \times C) \cap (\overline{B} \times C)$
(d) $A \triangle (B \times C) = (A \triangle B) \times (A \triangle C)$

(3) Find an example to show that union is not distributive with respect to the Cartesian product, *i.e.*, that the following is not an identity.

$$A \cup (B \times C) = (A \cup B) \times (A \cup C)$$

(4) Let A, B, C, and D be sets. Prove the following assertions.
(a) If $A \times A = B \times B$, then $A = B$
(b) If $A \subseteq B$, then $A \times C \subseteq B \times C$
(c) If $A \subseteq C$ and $B \subseteq D$, then $A \times B \subseteq C \times D$
(d) If $A \times B \subseteq A \times C$ and $A \neq \emptyset$, then $B \subseteq C$

(5) Does Definition 1.3.1 imply the following?

$$A \times B \times C = (A \times B) \times C = A \times (B \times C)$$

From the definition of the Cartesian product we have seen that any element (a,b) in a Cartesian product $A \times B$ is just an ordered pair. No relationship is required between the objects a and b for them to form an ordered pair. Thus, frequently we are not interested in the entire Cartesian product set, but only in a certain portion of it which is in some way well defined. As an illustration, consider Example 1.3.1. Suppose we are only interested in those ordered pairs whose second co-ordinate number is an integral multiple of their first co-ordinate number. Then we find that:

$$R_1 = \{(1,3),(1,4),(2,4)\} \subset A \times B$$

$$R_2 = \{(0,0),(1,0),(1,1),(1,2),(2,0),(2,2)\} \subset A \times A$$

Note that R_1 and R_2 are subsets of $A \times B$ and $A \times A$, respectively. We give the following definition.

Definition 1.3.2. A *(binary) relation R* from A to B is a subset of $A \times B$. If $A = B$, we say R is a (binary) relation in A. More generally, an *n-ary relation* is a subset of a Cartesian product on n sets A_1, $A_2, \ldots, A_n$.

Example 1.3.3. Consider the set P of a group of people. Suppose we are interested in those who are husband and wife in this group. The relation R in this example will be "husband and wife". Then:

$$R = \{(x,y) \mid x,y \in P \text{ and } x \text{ is the husband of } y \text{ or } y \text{ is the wife of } x\}$$

Obviously, R is a subset of $P \times P$, because some ordered pairs in $P \times P$, for example the ordered pairs (z,z) where z is a man or woman, are not in the relation "husband and wife".

Example 1.3.4. Suppose it is desired to find all the points inside the unit circle whose center is at the origin. Then the relation is

$$R = \{(x,y) \mid x \text{ and } y \text{ are real numbers and } x^2 + y^2 < 1\}$$

which is a relation in R^1.

Definition 1.3.3. Let R be a relation from A to B. The *domain* of R, denoted by Dom R, is defined:

$$\text{Dom } R = \{x \mid x \in A \text{ and } (x,y) \in R \text{ for some } y \in B\}$$

The *range* of R, denoted by Ran R, is defined:

$$\text{Ran } R = \{y \mid y \in B \text{ and } (x,y) \in R \text{ for some } x \in A\}$$

Clearly, Dom $R \subseteq A$ and Ran $R \subseteq B$. Moreover, the domain of R is the set of first co-ordinates in R and the range of R is the set of second co-ordinates in R. Thus, we can write the relation as follows.

$$R = \{(x,y) \mid x \in \text{Dom } R \text{ and } y \in \text{Ran } R\}$$

We sometimes write $(x,y) \in R$ as xRy which reads as "x relates y".

Definition 1.3.4. Let R be a relation in A. R is an *equivalence relation* in A iff the following conditions are satisfied.
(1) xRx for all $x \in A$ (R is reflexive)
(2) If xRy, then yRx, for all $x,y \in A$ (R is symmetric)
(3) If xRy and yRz, then xRz for all $x,y,z \in A$ (R is transitive)

Example 1.3.5. The identity relation in A, I_A, defined by

$$I_A = \{(x,y)\,|\,x \in A,\, y \in A \text{ and } x = y\}$$

is an equivalence relation in A.

Example 1.3.6. Let N be the set of natural numbers, that is, $N = \{1, 2, 3, \ldots\}$. Define a relation E in N as follows.

$$E = \{(x,y)\,|\,x,y \in N \text{ and } x + y \text{ is even}\}$$

E is an equivalence relation in N because the first two conditions are clearly satisfied. As to the third condition, if $x + y$ and $y + z$ are divisible by 2, then $x + (y + y) + z$ is divisible by 2. Hence $x + z$ is divisible by 2. In this equivalence relation all the odd numbers are equivalent and so are all the even numbers.

Exercises

(1) Find the fallacy in the following argument.

"If a relation R is symmetric and transitive it should follow that the relation is also reflexive.

Proof: Since the relation R is symmetric, then if aRb it follows that bRa. Since R is transitive, it follows that with aRb and bRa, aRa. Hence, R is also reflexive."

(2) What condition may we impose on a relation R in A in order to ensure that reflexivity is a consequence of symmetry and transitivity?

(3) State which of the following relations are equivalence relations.

For the set of all people:

(a) x is a parent of y

(b) x and y have a common ancestor

(c) x is a friend of y

(4) State, in each case, which of the three properties, i.e., (1) reflexivity, (2) symmetry, (3) transitivity, are possessed by the relations.

Let $A = \{1,2,3\}$. The following relations are defined in A.

(a) $R_1 = \{(2,2),(2,3),(3,2),(3,3)\}$

(b) $R_2 = \{(1,1),(2,2),(3,3),(1,2)\}$

(c) $R_3 = \{(1,1),(2,2),(1,2),(2,1)\}$

(d) $R_4 = \{(1,1),(2,2),(3,3),(1,2),(2,1),(2,3),(3,2)\}$

(5) Let $R = \{(x,y)\,|\,x \text{ and } y \text{ are real numbers and } y = x^2\}$. What are the domain and range of this relation?

We shall now turn our attention to an important class of relations called *functions*. The words map or mapping, transformation, correspondence, and operator are among those that are sometimes used as synonyms for function.

Definition 1.3.5. Let A and B be two nonempty sets. A *function*, denoted by f, from A to B is a relation from A to B such that:
(1) Dom $f = A$
(2) If $(x,y) \in f$ and $(x,z) \in f$, then $y = z$. This holds for all $x \in A$.
We write $f{:}A \to B$, which is read "f is a function from A to B". Function is also sometimes defined as follows. Let A and B be two nonempty sets. A function f from A to B is a rule which associates with each element in A an element in B.

Defintion 1.3.6. Let A and B be two nonempty sets and let $f{:}A \to B$. If $(x,y) \in f$, then we say that y is the *image* of x and write $y = f(x)$.

Definition 1.3.7. f is said to be *one-to-one* if $f(x_1) = y$ and $f(x_2) = y$ implies $x_1 = x_2$.

Definition 1.3.8. f is said to be a function from A *onto* B if Ran $f = B$.

Clearly, if f is a *one-to-one* and *onto* function, then there exists a one-to-one correspondence between sets A and B. If a function is not one-to-one, we call it a *many-to-one function*. If a function is not onto B, then it is *into B*.

The following question may arise. Is the function defined by Definition 1.3.5 the same as the functions we study in algebra, geometry, and calculus? The answer to this question is yes, provided that the function must be single valued according to Definition 1.3.5. When we write, for example, $y = 2x$, we really mean the following.

$$f = \{(x,y) \mid x \text{ and } y \text{ are real numbers and } y = 2x\}$$

It should also be noted that, since a function is a set, we naturally say that two functions are equal if and only if they are equal as sets.

Definition 1.3.9. If $f{:}A \to B$ is one-to-one and onto, then the inverse relation f^{-1} from B to A *will be called the inverse function of f.* (Notice that f^{-1} is single valued as required by the definition of a function.)

Exercises

(1) Let A and B be two nonempty sets. Show that if f is a one-to-one

and onto function from A to B, then f^{-1} is a one-to-one and onto function from B to A.

(2) If $A = \{1,2,3,4\}$ and $B = \{a,b,c,d\}$, which of the following are functions from A to B? If a set is such a function, give the range. If it is not, explain.

(a) $f_1 = \{(1,a),(2,b),(4,c),(3,d),(2,c)\}$
(b) $f_2 = \{(1,b),(2,c),(3,a),(4,d)\}$
(c) $f_3 = \{(1,a),(2,c),(3,b),(4,d)\}$

(3) Which of the following relations are functions from R' to R'. Give your reasons.

(a) $f_1 = \{(x,y)|x,y \in R' \text{ and } y = \sqrt{x}\}$
(b) $f_2 = \{(x,y)|x,y \in R' \text{ and } y = 1\}$
(c) $f_3 = \{(x,y)|x,y \in R' \text{ and } y^2 = x\}$

(4) Prove that if f is a one-to-one and onto function from A to B and if $X,Y \subseteq A$ and $W,Z \subseteq B$, then the following are true.

(a) $f(X \cup Y) = f(X) \cup f(Y)$
(b) $f^{-1}(W \cup Z) = f^{-1}(W) \cup f^{-1}(Z)$
(c) $f^{-1}(W \cap Z) = f^{-1}(W) \cap f^{-1}(Z)$
(d) $f^{-1}(B - W) = f^{-1}(B) - f^{-1}(W)$

(5) In Exercise 4, is it also true that $f(X \cap Y) = f(X) \cap f(Y)$? If not, can you tell what the relation between $f(X \cap Y)$ and $f(X) \cap f(Y)$ should be?

1.4. POWER SETS

As mentioned before, the elements of a set may themselves be sets. A special class of such sets is the power set.

Definition 1.4.1. Let A be a given set. The *power set* of A, denoted by $P(A)$, is a family of sets such that if $X \subseteq A$, then $X \in P(A)$. Symbolically, $P(A) = \{X|X \subseteq A\}$.

Example 1.4.1. The power set of the empty set $\varnothing$ is a singleton $\{\varnothing\}$.

Example 1.4.2. Let $A = \{a,b,c\}$. The power set of A is as follows.

$$P(A) = \left\{ \{\varnothing\},\{a\},\{b\},\{c\},\{a,b\},\{b,c\},\{c,a\},\{a,b,c\} \right\}$$

The following is an important property of the power set.

Theorem 1.4.1. Prove that if a set A has exactly n elements, then P(A) will have exactly 2^n elements.

Proof: Each element of A is either in or not in some subset of A. Thus there are n independent binary choices or 2^n ways to choose a subset. In other words, there must be 2^n elements in $P(A)$.

Exercises

(1) Prove that if a set A has exactly n elements, then $P(\underbrace{P \ldots (P(A)) \ldots)}_{m}$ will have exactly $\underbrace{2^{2 \cdot \cdot \cdot^{2^n}}}_{m}$ elements.

(2) Let A and B be two sets. Prove the following.

$$P(A) \cap P(B) = P(A \cap B)$$

(3) Is the following equation true?

$$P(A) \cup P(B) = P(A \cup B)$$

(4) State which of the following relations are equivalence relations. For the power set $P(A)$ of a given set A:
   (a) $X \subset Y$
   (b) $X \cap Y = \emptyset$
   (c) $X \cap Y \neq \emptyset$
   (d) $X \cup Y = A$

1.5. CONVEX SETS

In this section we shall introduce another class of sets, called *convex sets*, which has remarkably widespread applications in both pure and applied mathematics.

Definition 1.5.1. Let S be a real linear space. A set A in S is called *convex* if when $x_1, x_2 \in A$, then also $\alpha_1 x_1 + \alpha_2 x_2 \in A$ whenever α_1 and α_2 are positive numbers such that $\alpha_1 + \alpha_2 = 1$.

Definition 1.5.2. Let x_1 and x_2 be two points of a real linear space S. The set of all points $\alpha x_1 + (1 - \alpha)x_2$ for which $0 \leq \alpha \leq 1$ is called the *line segment* joining x_1 and x_2, or simply the line segment $\overline{x_1 x_2}$.

Definition 1.5.3. A set A in a real linear space S is called *convex* if A contains the line segment $x_1 x_2$ whenever x_1 and x_2 are two points of A.

Example 1.5.1. The empty set and a set containing a single element are convex.

Example 1.5.2. A set M of vectors in S is called *linear manifold* if, for all scalars α and β, it contains the vectors $\alpha x + \beta y$ whenever it contains the vectors x and y. A linear *manifold* is convex.

Example 1.5.3. Let S be a real linear space, and A and B be convex sets in S. Let α and β be any scalars. Define $\alpha A + \beta B$ as follows.

$$\alpha A + \beta B = \{z \mid z = \alpha x + \beta y, \text{ where } x \in A \text{ and } y \in B\}$$

Then the set $\alpha A + \beta B$ is convex.

Theorem 1.5.1. Let A and B be convex sets. Then $A \cap B$ is convex. More generally, if $A_1, A_2, \ldots, A_n$ are convex, then $\bigcap_{i=1}^{n} A_i$ is convex.

Proof. If $A \cap B$ is empty or a singleton, then it is convex. Suppose $A \cap B$ has more than one element. Let x and y be any two distinct points in the intersection of A and B. Since A and B are convex, both A and B contain the line segment $\overline{xy}$. Thus the intersection $A \cap B$ contains the line segment $\overline{xy}$.

Definition 1.5.4. Let S be a linear space and x be a vector in S. A *linear functional* defined on S, $f(x)$, is a mapping of the linear space S into a one-dimensional space such that

$$f(\alpha x + \beta y) = \alpha f(x) + \beta f(y)$$

where x, y are any two vectors and α, β are scalars.

A set of vectors defined by

$$H_\lambda^o = \{x \mid f(x) < \lambda, \text{ where } \lambda \text{ is a fixed scalar}\}$$

is called an open half-space, and the set $H_\lambda^c = \{x \mid f(x) \le \lambda\}$ is called a closed half-space.

Theorem 1.5.2. A half-space (open or closed) is convex.

Proof. Let H_λ^o be an open half-space, and x and y be two vectors in H_λ^o. Then $f(x) < \lambda$ and $f(y) < \lambda$. Let $0 \le \alpha \le 1$. Then:

$$f(\alpha x + (1 - \alpha)y) = \alpha f(x) + (1 - \alpha)f(y) < \alpha \omega + (1 - \alpha)\omega = \omega$$

Thus, H_λ^o is convex. That H_λ^c is convex can be proven similarly.

From Theorems 1.5.1 and 1.5.2 we immediately have the following consequence.

Theorem 1.5.3. The intersection of half-spaces is a convex set.
We omit the proof.

Now consider the intersection of a finite number of closed half-spaces.

$$H^c_{\lambda_i} = \{x \,|\, f_i(x) \leqslant \lambda_i\}, i = 1, 2, \ldots, n$$

Let $C = \bigcap_{i=1}^{n} H^c_{\lambda_i}$, which is called a *polyhedral convex set*. The sets B_i, defined by $B_i = \{x f_i(x) = \lambda_i\}$, are called the *bounding hyperplanes* of the half-space $H^c_{\lambda_i}$. The *extreme points* of C are defined as the intersections of the n bounding hyperplanes of C. Furthermore, we define the convex combination of n vectors as follows.

Definition 1.5.5. Let $x_1, x_2, \ldots, x_n$ be n vectors in S and $\alpha_1, a_2, \ldots,$ α_n be non-negative numbers such that $\alpha_1 + \alpha_2 + \ldots + \alpha_n = 1$. The vector y formulated by

$$y = \alpha_1 x_1 + \alpha_2 x_2 + \ldots + \alpha_n x_n$$

is called a *convex combination* of $x_1, x_2, \ldots, x_n$. Then we have the following.

Theorem 1.5.4. Let

$$C = \bigcap_{i=1}^{n} H^c_{\lambda_i}$$

be a bounded polyhedral convex set and $x_1, x_2, \ldots, x_n$ be all its extreme points.

  (a) Every convex combination of $x_1, x_2, \ldots, x_n$ belongs to C.
  (b) Every vector of C is expressible as a convex combination of $x_1, x_2, \ldots, x_n$.

Proof
(a) Let y be a convex combination of $x_1, x_2, \ldots, x_n$. That is,

$$y = \alpha_1 x_1 + \alpha_2 x_2 + \ldots + \alpha_n x_n$$

where the α_i's are non-negative numbers and $\alpha_1 + \alpha_2 + \ldots + \alpha_n = 1$. Since $x_1, x_2, \ldots, x_n$ are extreme points of C, $f_i(x_k) \leqslant \lambda_i, i = 1, 2, \ldots, n$, implies that

$$x_k \in C = \bigcap_{i=1}^{n} \{x \,|\, f_i(x) \leqslant \lambda_i\}$$

for $k = 1, 2, \ldots, n$. Since the f_i are linear functionals,

$$f_i(\alpha_k x_k) = \alpha_k f_i(x_k) \leqslant \alpha_k \lambda_i, i = 1, 2, \ldots, n,$$

implies that:

$$\alpha_k x_k \in \alpha_k C = \bigcap_{i=1}^{n} \{\alpha_k x \mid f_i(\alpha_k x) \leqslant \alpha_k \lambda_i\}.$$

Then

$$f_i(y) = f_i\left(\sum_{k=1}^{n} \alpha_k x_k\right) = \sum_{k=1}^{n} f_i(\alpha_k x_k) = \sum_{k=1}^{n} \alpha_k f_i(x_k)$$
$$\leqslant \left(\sum_{k=1}^{n} \alpha_k\right)\lambda_i, \quad i = 1, 2, \ldots, n$$

implies that

$$y \in \left(\sum_{k=1}^{n} \alpha_k\right) C = C.$$

Hence the proof.

(b) Every segment connecting two adjacent extreme points x_i and x_{i+1} (see Fig. 1.1) may be expressed as $x_i x_{i+1} - \alpha x_i + (1 - \alpha) x_{i+1}$, where $0 \leqslant \alpha \leqslant 1$. Let y be a point in C. Connect it with any point of the segment, say z_1.

$$z_1 = \alpha_0 x_i + (1 - \alpha_0) x_{i+1}$$

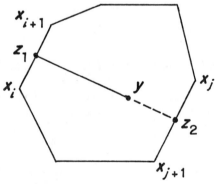

FIG. 1.1.   Proof of Theorem 1.5.4(*b*)

Then extend the line segment $\overline{z_1 y}$ until it intersects with another bounding hyperplane at the other end at, say, z_2, which is on the line segment $\overline{x_j x_{j+1}}$. That is, for some $0 \leqslant \beta_0 \leqslant 1$

$$z_2 = \beta_0 x_j + (1 - \beta_0) x_{j+1}$$

The line segment $\overline{z_1 z_2}$ may be represented by

$$\overline{z_1 z_2} = \gamma z_1 + (1 - \gamma)z_2$$

where $0 \leqslant \gamma \leqslant 1$. Since y is on this line segment, there exists γ_0, $0 \leqslant \gamma_0 \leqslant 1$, such that:

$$\begin{aligned}
y &= \gamma_0 z_1 + (1 - \gamma_0)z_2 \\
&= \gamma \left[\alpha_0 x_i + (1 - \alpha_0)x_{i+1}\right] + (1 - \gamma_0)\left[\beta_0 x_j + (1 - \beta_0 x_{j+1}\right] \\
&= \gamma_0 \alpha_0 x_i + \gamma_0(1 - \alpha_0)x_{i+1} + (1 - \gamma_0)\beta_0 x_j + (1 - \gamma_0)(1 - \beta_0)x_{j+1}
\end{aligned}$$

We observe that the algebraic sum of the coefficients of x_i, x_{i+1}, x_j, and x_{j+1} is as follows:

$$\gamma_0 \alpha_0 + \gamma_0(1 - \alpha_0) + (1 - \gamma_0)\beta_0 + (1 - \gamma_0)(1 - \beta_0) = 1$$

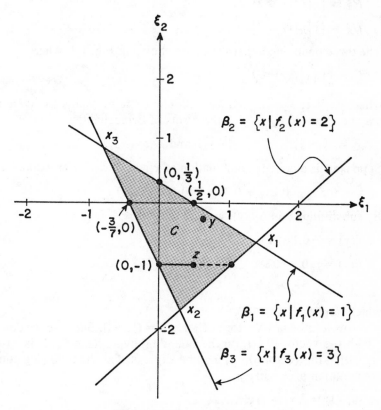

FIG. 1.2. Example 1.5.4

Thus y is a convex combination of $x_1, x_2, \ldots, x_n$. This completes the proof.

Example 1.5.4. Let $x = (\xi_1, \xi_2)$ be a vector in a two-dimensional Euclidean space S and $f_1(x)$, $f_2(x)$, and $f_3(x)$ be three linear functionals defined as follows.

$$f_1(x): = \quad 2\xi_1 + 3\xi_2$$

$$f_2(x): = \quad \xi_1 - \xi_2$$

$$f_3(x): = -7\xi_1 - 3\xi_2$$

Let H_1^c, H_2^c, and H_3^c be three closed half-spaces defined by:

$$H_1^c = \{x \mid f_1(x) \leq 1\}$$

$$H_2^c = \{x \mid f_2(x) \leq 2\}$$

$$H_3^c = \{x \mid f_3(x) \leq 3\}$$

The three bounding hyperplanes are shown in Fig. 1.2, where

$$C = \bigcap_{i=1}^{3} \{x \mid f_i(x) \leq \lambda_i\}$$

where $\lambda_1 = 1$, $\lambda_2 = 2$, and $\lambda_3 = 3$. The extreme points x_1, x_2, and x_3 may be obtained by solving three sets of two equalities:

$$x_1 = (\tfrac{7}{5}, -\tfrac{3}{5}), \quad x_2 = (\tfrac{3}{10}, -\tfrac{17}{10}), \text{ and } x_3 = (-\tfrac{4}{5}, \tfrac{13}{15})$$

Suppose y is a convex combination of x_1, x_2, and x_3. For example:

$$y = 0.5x_1 + 0.2x_2 + 0.3x_3 = (0.52, -0.38)$$

By substituting y into $f_i(x)$, we find that

$$f_1(y) = -0.1 < 1$$

$$f_2(y) = \cdot \ 0.9 < 2$$

$$f_3(y) = -2.5 < 3$$

which implies that $y \in C$.

Now let z be any vector in C, say $z = (\tfrac{1}{2}, -1)$. Suppose we connect z with the vector $(0, -1)$ on the bounding hyperplane β_3. It is easy to show that the vector $(0, -1)$ may be expressed as the following convex combination of x_2 and x_3:

$$(0, -1) = \tfrac{8}{11} x_2 + \tfrac{3}{11} x_3$$

If we extend this line, it will intersect β_2 at the point $(1,-1)$, which may be expressed as the convex combination of x_1 and x_2:

$$(1,-1) = \tfrac{7}{11}x_1 + \tfrac{4}{11}x_2$$

Furthermore, we find the vector z may also be expressed as the convex combination of the vectors $(0,-1)$ and $(1,-1)$:

$$z = \tfrac{1}{2}(0,-1) + \tfrac{1}{2}(1,-1)$$

Substitution of the above convex combination expressions for $(0,-1)$ and $(1,-1)$ yields:

$$z = \tfrac{7}{22}x_1 + \tfrac{12}{22}x_2 + \tfrac{3}{22}x_3$$

Hence z is expressible as a convex combination of x_1, x_2, and x_3. It is clear that this expression is not unique; in fact we may obtain infinitely many such expressions for a given z in C.

Exercises

(1) Let H^o be an open half-space in E^2. H^o is defined by:

$$H^o = \{x \mid f(x) < 1, \text{ where } f(x) = \xi_1 + 2\xi_2 \text{ and } x = (\xi_1,\xi_2)\}$$

Define the space $H^{oc} = \{x \mid f(x) = 1\}$. Let $x_1 = (\tfrac{1}{2},0)$ and $x_2 = (0,\tfrac{1}{2})$ be two vectors in H^o and H^{oc}, respectively, and $y = 0.75x_1 + 0.25x_2$. Show that y is in H^o.

(2) The result of Problem 1 may be generalized as follows. Let A be an open convex set in a linear space S. Define the closure of A, denoted by A^c, to be the intersection of all closed subsets of S which contain A. Then every point y expressible in the form

$$y = \alpha x_1 + (1 - \alpha)x_2$$

where $x_1 \in A$ and $x_2 \in A^c$, is in A.

(3) Define four linear functionals f_1, f_2, f_3, and f_4, on E^3 as

$$f_1(x) = \xi_1 + 2\xi_2 + 2\xi_3 + 4\xi_4$$
$$f_2(x) = 4\xi_1 + 2\xi_2 - 2\xi_3 - 3\xi_3$$
$$f_3(x) = 5\xi_1 - \xi_2$$
$$f_4(x) = \xi_3 - \xi_4$$

where $x = (\xi_1, \xi_2, \xi_3, \xi_4)$, and define four closed half-planes as

$$H_i^c = \{x \mid f_i(x) \leq \lambda_1\}$$

where $\lambda_1 = -1$, $\lambda_2 = 1$, $\lambda_3 = 4$, and $\lambda_4 = 0$.

(a) Find the extreme points x_1, x_2, x_3, and x_4 of the polyhedral convex set $C = \bigcap_{i=1}^{4} H_i^c$.

(b) Prove that $y = 0.1x_1 + 0.2x_2 + 0.3x_3 + 0.4x_4$ belongs to C.

(c) Let $z = (1,1,-1,-1)$. Find a set of α_1, α_2, α_3, and α_4 such that

$$z = \alpha_1 x_1 + \alpha_2 x_2 + \alpha_3 x_3 + \alpha_4 x_4$$

where all α_i are positive and

$$\sum_{i=1}^{4} \alpha_i = 1.$$

1.6. LATTICE AND ITS BASIC PROPERTIES

We begin by introducing the partial ordering relation. We define the relation $x \leq y$ as "y includes x" and the relation $x < y$ as "y strictly includes x".

Definition 1.6.1. A relation $\leq$ on a set A is said to be a *partial ordering* in A if it satisfies the following axioms.

(1) Reflexive: for all $x \in A$, $x \leq x$

(2) Antisymmetric: if $x,y \in A$, $x \leq y$ and $y \leq x$, then $x = y$

(3) Transitive: if $x,y,z \in A$, $x \leq y$ and $y \leq z$, then $x \leq z$

A set P over which a relation $\leq$ of partial ordering is defined is called a *partially ordered* set or a *poset*.

Definition 1.6.2. An element m in P is a *meet* of x and y if it is a g.l.b. (greatest lower bound) of x and y. An element l in P is a *join* of x and y if it is an l.u.b. (least upper bound) of x and y. We shall denote the meet and join of x and y by $m = x \cap y$ and $l = x \cup y$, respectively.

Here, we are mainly interested in the algebraic aspects of ordered sets. We use $\cdot$ and $+$ to represent the meet and the join operations. We shall use x,y,z, to denote generic elements in an ordered set and a,b,c to denote specific elements. The analogous quantities among sets, ordered sets, and algebras are shown in Table 1.6.1.

From the definitions of the inclusion relation and the meet and join operations, we have the following.

Lemma 1.6.1. In any algebraic poset A the meet and join of two elements of A, if they exist, have the following property:

$$x \leqslant y \text{ iff } x \cdot y = x \text{ and } x + y = y$$

This property is often referred to as the *consistency* property.

Proof. The proof is obvious and can thus be omitted.

Table 1.2.1. The analogous quantities among sets, ordered sets, and algebras

Quantity	Sets	Ordered sets	Algebras
Relation	$\subseteq$: containment $X \subseteq Y$ means that Y contains X	$\leqslant$: inclusion relation $x \leqslant y$ means that y includes x	$\leqslant$: inequality $a \leqslant b$ means that b is equal to or greater than a
Operation	$\cup$: union $X \cup Y$ means the set whose elements are in X, or in Y, or in both $\cap$: intersection $X \cap Y$ means the set whose elements are in both X and Y	$\cup$: join $x \cup y$ means the l.u.b. of x and y $\cap$: meet $x \cap y$ means the g.l.b. of x and y	+ : l.u.b. $a + b$ means the l.u.b. of a and b $\cdot$: g.l.b. $a \cdot b$ means the g.l.b. of a and b
Element	A The universal set $\varnothing$ The empty set	I The greatest element 0 The least element	1 The largest number 0 The smallest number

The following theorem states the properties of the meet and join in a poset.

Theorem 1.6.1. In any poset A the meet and join operations of two

elements A, if they exist, satisfy the idempotent, commutative, associative, and absorptive properties, i.e., for all x,y,z in A

(1) *idempotent:*   L1. : $x \cdot x = x$ L1$_+$: $x + x = x$

(2) *commutative:*   L2. : $x \cdot y = y \cdot x$ L2$_+$: $x + y = y + x$

(3) *associative:*   L3. : $x \cdot (y \cdot z)$ L3$_+$: $x + (y + z) = (x + y) + z$

                                 $= (x \cdot y) \cdot z$

(4) *absorptive:*   L4. : $= x \cdot (x + y) = x$ L4$_+$: $x + (x \cdot y) = x$

Proof. The properties L1 and L2 follow directly from the definitions of meet and join. The properties L3 are evident since $x \cdot (y+z)$ and $(x \cdot y) + z$ are both equal to the g.l.b. of x, y, and z, and $x + (y+z)$ and $(x+y) + z$ are both equal to the l.u.b. of x, y, and z. To prove L4, consider the following two cases.

(i) If $x \leqslant y$, then

$$x \cdot (x+y) \;\; = x \cdot y \qquad \text{by Lemma 1.6.1}$$

$$= x \qquad \text{by Lemma 1.6.1}$$

and

$$x + (x \cdot y) = x + x \qquad \text{by Lemma 1.6.1}$$

$$= x \qquad \text{by L1}_+$$

(ii) If $y \leqslant x$, then

$$x \cdot (x+y) \;\; = x \cdot (y+x) \quad \text{by L2}_+$$

$$= x \cdot x \qquad \text{by Lemma 1.6.1}$$

$$= x \qquad \text{by L1.}$$

and

$$x + (x \cdot y) = x + (y \cdot x) \quad \text{by L2.}$$

$$= x + y \qquad \text{by Lemma 1.6.1}$$

$$= x \qquad \text{by Lemma 1.6.1.}$$

Hence the meet and join operations also satisfy the absorption property. A lattice is defined as follows.

Definition 1.6.3. A lattice is a poset L in which any two elements x and y have both a meet and a join.

A lattice has the following properties.

Theorem 1.6.2. In any lattice the following hold.

(1) *All the elements satisfy L1–L4 of Theorem 1.6.1.*

(2) *All elements satisfy the isotone property, that is, if $x \leqslant y$, then $x \cdot z \leqslant y \cdot z$ and $x + z \leqslant y + z$.*

(3) *All elements satisfy the modular inequality, which is, if $x \leqslant z$, then $x + (y \cdot z) \leqslant (x+y) \cdot z$.*

(4) *The distributive inequalities are satisfied:*

$$x \cdot (y+z) \geqslant (x \cdot y) + (x \cdot z)$$

$$x+(y \cdot z) \leqslant (x+y) \cdot (x+z)$$

Proof. Property (1) is evident from the definition of lattice. Properties (2), (3), and (4) may be proven by the following algebra.

Proof of property 2. If $x \leqslant y$, then

$$x \cdot z = (x \cdot y) \cdot (z \cdot z) \quad \text{by Lemma 1.6.1 and L1}$$

$$= (x \cdot z) \cdot (y \cdot z) \quad \text{by L2}$$

which implies $(x \cdot z) \leqslant (y \cdot z)$ by Lemma 1.6.1. The second inequality may be proven using the duality principle.

Proof of property 3. Since $x \leqslant z$ and $x \leqslant x + y$,

$$x \leqslant (x+y) \cdot z$$

and since $y \cdot z \leqslant z$ and $y \cdot z \leqslant y \leqslant x + y$

$$y \cdot z \leqslant (x+y) \cdot z$$

Combining these results and in view of the definition of $+$, we obtain

$$x + (y \cdot z) \leqslant (x+y) \cdot z.$$

Proof of property 4. The proof of this property is similar to that of property 3. Since $x \cdot y \leqslant x$ and $x \cdot y \leqslant y \leqslant y + z$,

$$x \cdot y \leqslant x \cdot (y+z).$$

From the relations $x \cdot z \leqslant x$ and $x \cdot z \leqslant z \leqslant y + z$:

$$x \cdot z \leqslant x \cdot (y+z)$$

Hence

$$x \cdot (y+z) \geqslant (x \cdot y) + (x \cdot z)$$

Again, the second inequality may be proven using the duality principle.

Theorem 1.6.3. Every finite lattice has a least element and a greatest element.

Proof. Let the elements of a finite lattice L be $x_1, x_2, \ldots, x_n$. The least element of L is the element $x_1 \cdot x_2 \cdot \ldots \cdot x_n$, and the greatest element of L is the element $x_1 + x_2 + \ldots + x_n$.

Example 1.6.1. Consider the set L of the positive integral divisors of the natural number 216, i.e.,

$$L = \{1,2,3,4,6,8,9,12,18,24,27,36,54,72,108,216\}$$

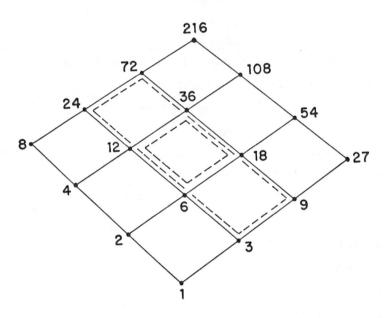

FIG. 1.3. The inclusion diagram of L of Example 1.6.1.

The inclusion diagram of L is as shown in Fig. 1.3. The multiplication (g.l.b.) and addition (l.u.b.) of any two elements of L may be found directly from the diagram. For instance, the meet and join of 12 and 18 are 6 and 36. The meet and join of 24 and 9 are 3 and 72. The meet and join of 4 and 108 are 4 and 108. It is seen that any two elements have a meet and a join. The inclusion diagram provides a convenient means of finding them. Hence L is a lattice. One may easily verify that all the properties of Theorems 1.6.1 and 1.6.2 are satisfied.

Now we want to prove that any non-empty set satisfying axioms L1–L4 is a lattice. First we define the following.

Definition 1.6.4. A non-empty set with a single binary operation which satisfies L1–L3 is called a *semi-lattice*.

Lemma 1.6.2. If P is a poset in which any two elements have a meet (join), P is a semi-lattice and is called a meet (join) lattice. Conversely, we have the following lemma.

Lemma 1.6.3. Let S be a semi-lattice under a binary operation $\circ$, and x,y be two elements of S:

(a) *the definition*

$$x \leq y \text{ iff } x \circ y = x$$

makes S a poset in which $x \circ y =$ g.l.b. $\{x,y\}$, i.e., $\circ$ is a meet operation on S, and

(b) *the definition*

$$x \leq y \text{ iff } x \circ y = y$$

makes S a poset in which $x \circ y =$ l.u.b. $\{x,y\}$, i.e., $\circ$ is a join operation on S.

Proof. The proof of (a) contains two parts. First, we want to prove that the relation $\circ$ is a partial ordering in A, and then to show that, if $m \leq x$ and $m \leq y$, $b \leq m$ for all b such that $b \leq x$ and $b \leq y$.

To show that the relation $\circ$ is a partial ordering in A is to show that the relation $\circ$ satisfies the reflexive law, antisymmetric law, and transitive law. By the definition of semi-lattice, the relation $\circ$ is idempotent, commutative, and associative. The idempotent law $x \circ x = x$ implies the reflexive law $x \leq x$. If $x \leq y$ (iff $x \circ y = x$) and $y \leq x$ (iff $y \circ x = y$), by the commutative law, i.e.,

$$x \circ y = y \circ x, \quad x = x \circ y = y \circ x = y$$

This proves that the relation $\circ$ satisfies the antisymmetric law. By the associative law $x \leq y$ (iff $x \circ y = x$) and $y \leq z$ (iff $y \circ z = y$) implies

$$x = x \circ y = x \circ (y \circ z) = (x \circ y) \circ z = x \circ z$$

i.e., $x \leq z$. This shows that the transitive law is also satisfied by the relation $\circ$. Hence $\circ$ is a partial ordering relation in S.

Next, we want to show that $\circ$ is a meet operation in S,

$$(x \circ y) \circ x = x \circ (x \circ y)$$

$$= (x \circ x) \circ y$$

$$= x \circ y$$

$$(x \circ y) \circ x = x \circ y \Leftrightarrow x \circ y \leq x.$$

Similarly, we can show $x \circ y \leq y$. Now if $b \leq x$ and $b \leq y$, then

$$b \circ (x \circ y) = (b \circ x) \circ z = b \circ y = b$$

which implies $b \leq x \circ y$. This proves that $x \circ y =$ g.l.b. $\{x,y\}$. The proof of (b) may be obtained similarly. Hence the theorem.

Lemma 1.6.4. Let S be a non-empty set with the multiplication and addition operations defined on it. If the multiplication and addition operations satisfy the absorption properties L4. and L4$_+$, then:

$$x \leq y \Leftrightarrow x \cdot y = x \text{ and } x \leq y \Leftrightarrow x + y = y$$

Proof. Let x,y be two elements of S. Since the multiplication and addition operations satisfy L4$_+$, $x \cdot y = x$ implies

$$x + y = (x \cdot y) + y = y$$

and $x + y = y$ implies

$$x \cdot y = x \cdot (x+y) = x$$

Conversely, if $x \leq y$,

$$x = x \cdot (x+y) = x \cdot y \quad \text{by L4.}$$

$$y = x + (x \cdot y) = x + y \text{ by L4}_+$$

Thus $x \leq y \Rightarrow x \cdot y = x$ and $x \leq y \Rightarrow x + y = y$. Hence the lemma.

From Lemmas 1.6.3 and 1.6.4 we immediately have the following theorem.

Theorem 1.6.4. Any non-empty set L with two binary operations which satisfies L1–L4 is a lattice.

Proof. The proof is evident from Lemmas 1.6.3 and 1.6.4. Similarly to the poset, a lattice may also be a Cartesian product of two sets.

Theorem 1.6.5. The Cartesian product of two lattices A and B with the inclusion relation defined as

$$(a_1,b_1) \leq (a_2,b_2) \text{ iff } a_1 \leq a_2 \text{ in } A \text{ and } b_1 \leq b_2 \text{ in } B$$

is a lattice. More generally, the Cartesian product $A_1 \times A_2 \times \ldots \times A_n$, where $A_1, A_2, \ldots, A_n$ are lattices, with the inclusion relation defined as

$$(a_1,a_2, \ldots, a_n) \leq (a_1',a_2', \ldots, a_n') \text{ iff } a_1 \leq a_1' \text{ in } A_1$$

$$a_2 \leq a_2' \text{ in } A_2, \ldots, a_n \leq a_n' \text{ in } A_n$$

is a lattice.

Proof. Again, the proof is evident.

Two interesting examples of Cartesian product lattices of two special classes are given below. The first is a *Boolean lattice,* and the second is a lattice called the *factorization lattice.*

Example 1.6.2. Let us consider two lattices $A = \{1,2,4,8\}$ and $B = \{1,2,3,6\}$, both relative to the positive integral divisibility relation. The Cartesian product $A \times B$ is shown in Fig. 1.4. Obviously, it is a lattice.

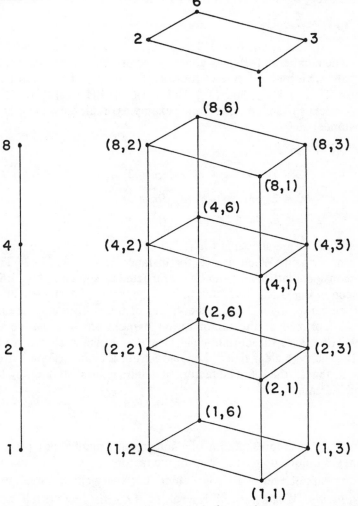

FIG. 1.4. Cartesian product of $A = \{1,2,4,8\}$ and $B = \{1,2,3,6\}$

Example 1.6.3. We shall consider an example of a factorization lattice. Let L be the set of *all* the factors of some natural number

$$x = p_1^{m_1} \times p_2^{m_2} \times \ldots \times p_n^{m_n}$$

where the p_i are distinct primes. From elementary number theory it can be shown that, for any integer $x > 1$, the representation of x as a product of primes is unique up to the order of the factors. The proof of this fact is left to the reader as an exercise.

Since every member of L is of the form

$$p_1^\alpha \times p_2^\beta \times \ldots \times p_n^\nu$$

where $\alpha = 0, 1, \ldots, m_1$, $\beta = 0, 1, \ldots, m_2$, $\ldots$, $\nu = 0, 1, \ldots, m_n$, we see that the number of factors is given by the product of the exponents, each of which has been increased by 1; for instance, $113400 = 2^3 \cdot 3^4 \cdot 5^2 \cdot 7$ has $(3 + 1)(4 + 1)(2 + 1)(1 + 1) = 120$ factors. Now let us examine some simple examples of this class of lattices. For instance,

L_1: $x = 4 = 2^2$ $(n = 1)$

L_2: $x = 12 = 2^2 \cdot 3$ $(n = 2)$

L_3: $x = 60 = 2^2 \cdot 3 \cdot 5$ $(n = 3)$

L_4: $x = 2940 = 2^2 \cdot 3 \cdot 5 \cdot 7^2$ $(n = 4)$

The four lattices are shown in Fig. 1.5. In Fig. 1.5*a*, $n = 1$, the factors of 4 form a chain. When $n = 2$, for instance, $x = 12 = 2^2 \cdot 3$. This two-dimensional lattice (L_2) can be constructed as shown in Fig. 1.5*b*. As is seen in Fig. 1.5*c*, the factorization expression of each element of the lattice can be by a 3-tuple; the values of the co-ordinates are the powers of primes of the factorization expression. The lattice for $x = 2940$ (there are 36 factors), which is an $n = 4$ case, is shown in Fig. 1.5*d*. It should be clear that we can construct a k-dimensional lattice for $n = k$, that is, the set of all factors of a finite natural number, by the method indicated.

Exercises

(1) Prove that the representation of a finite natural number as a product of primes is unique up to the order of the factors.

(2) Show that not every lattice is a chain, but every chain is a lattice.

(3) A *sublattice* M of a lattice L is a subset of L satisfying the following

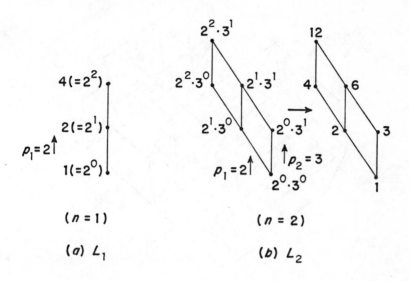

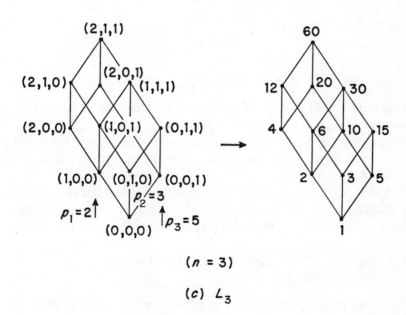

FIG. 1.5(*a, b, c*). Factorization lattices of Example 1.6.3

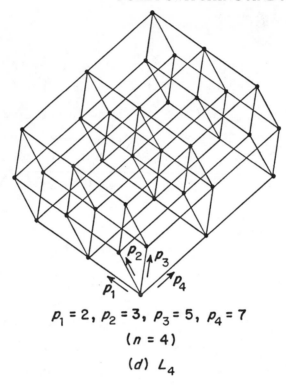

$$p_1 = 2, \ p_2 = 3, \ p_3 = 5, \ p_4 = 7$$

$$(n = 4)$$

$$(d) \ L_4$$

FIG. 1.5d

condition: $x \in M \Rightarrow xy \in M$ and $xvy \in M$. Show that a lattice L is a chain iff every subset of L is a sublattice.

(4) A lattice L is called *complete* if every subset of L has an l.u.b. and a g.l.b. Prove that a lattice is complete unless it has a subset which forms an infinite chain.

(5) Let L_1, L_2 be two lattices. Show that the product lattice $L_1 \times L_2$ is isomorphic to the product lattice $L_2 \times L_1$. Also show that the projection $L_1 \times L_2 \rightarrow L_1$ is homomorphic to the projection $L_1 \times L_2 \rightarrow L_2$.

1.7. BOOLEAN ALGEBRA AND ITS BASIC PROPERTIES

There are many special classes of lattice, among which modular lattices, distributive lattices, and Boolean algebra (complemented distributive lattices or Boolean lattices) are the most important.

Definition 1.7.1. A lattice L is *modular* if it satisfies the following modular identity. For all x, y, z in L

L5:   If $x \leqslant z$, then $x + (y \cdot z) = (x + y) \cdot z$

Definition 1.7.2. A lattice L is *distributive* if it satisfies the distributive equalities

L6a: $x \cdot (y + z) = (x \cdot y) + (x \cdot z)$

L6b: $x + (y \cdot z) = (x + y) \cdot (x + z)$

Note that every distributive lattice is a modular lattice.

A special class of distributive lattice is the *complemented distributive* lattice or the *Boolean lattice*.

Definition 1.7.3. Let L be a lattice having both a zero element (least element) 0 and a unit element (greatest element) 1. Let x, y be elements of L. If $x \cdot y = 0$ and $x + y = 1$, then y is a complement of x and x is a complement of y. Complement such as the element 12 of the lattice in Fig. 1.3, may have several complements such as the element a of the lattice M_5 in Fig. 1.6a which has two complements (b and c), or the element b of the pentagonal lattice in Fig. 1.6b which has two complements (a and c) or may have a unique element such as the element a (or c) of the lattice of Fig. 1.6b which has a unique complement b.

From Theorem 1.6.3 and the definitions of 0 and 1, 0 is the *unique* complement of 1 and 1 is the unique complement of 0.

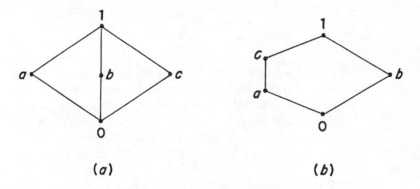

(a)   (b)

FIG. 1.6.   Two Boolean lattices

Definition 1.7.4. A lattice L is called *complemented* if all its elements have complements.

Example 1.7.1. Consider the power set $P(I)$ of all subsets S of a set I. The set $(P(I), \subseteq)$ with the meet and join operations defined by the set-theoretic intersection and set-theoretic union is a lattice. Since the usual set-theoretic complement $\overline{S}$ of S in I does satisfy the relations $\overline{S} \cap S = \varnothing$ (zero element) and $\overline{S} \cup S = I$ (unit element). Therefore the algebraic system $(P(I), \subseteq, \cap, \cup)$ is a Boolean lattice.

In this example, we have seen that the complement of every element of $P(I)$ is unique. In fact, every Boolean lattice has such property. This is stated in the following theorem.

Theorem 1.7.1. In a Boolean lattice each element has one and only one complement.

Proof. Let x be an element of a Boolean lattice. Suppose x has two complements y and z. Then

$$x \cdot y = x \cdot z = 0$$
$$x + y = x + z = 1$$

It follows that

$$
\begin{aligned}
y &= y \cdot (y + z) && \text{by L4} \\
&= y \cdot (x + z) && \text{by hypothesis} \\
&= z + (x \cdot y) && \text{by L5} \\
&= z + (x \cdot z) && \text{by hypothesis} \\
&= z && \text{by L4}_+
\end{aligned}
$$

Definition 1.7.5. A Boolean lattice considered as an algebra closed with respect to the three operations of complementation, formation of meet, formation of join, is called a *Boolean algebra*. Conventionally, in Boolean algebras we call the operations $\cdot$ and $+$ *multiplication* and *addition* respectively, instead of meet and join. The element $x \cdot y$ is called the *product* of x and y, and the element $x + y$ is called the *sum* of x and y.

An alternative definition of Boolean algebra may be defined as follows.

Definition 1.7.6. A *Boolean algebra* is an algebraic system $(B; +, \cdot, '; 0,1)$ consisting of a set B, binary operations $+$ and $\cdot$, and a unary operation $'$ satisfying axioms L1–L4, L6, and the following postulate.

L7: For every element x in B, there is a complement x' such that $x \cdot x' = 0$ and $x + x' = 1$. The zero and unit elements in a Boolean algebra are called degenerate if its multiplicative identity is equal to its additive identity, i.e., if $0 = 1$.

A simplest example of such an algebra is the one which is isomorphic to the power set $P(\varnothing)$ of the empty set, $P(\varnothing) = \{0\}$. In future we shall consider *only* non-degenerate algebras; i.e., Boolean algebras in which 0 and 1 are distinct. As a consequence of Theorem 1.7.1 we have the following.

Corollary 1.7.1. Any (non-degenerate) Boolean algebra must contain an even number of elements.

Example 1.7.2. Consider the (simplest non-degenerate) two-element Boolean algebra $B_2 = \{0,1\}$; 0 and 1 are complements of each other. From the definitions of meet (product) and join (sum), we can easily construct the following multiplication, addition, and complementation tables.

$\cdot$	0	1
0	0	0
1	0	1

$+$	0	1
0	0	1
1	1	1

$'$	
0	1
1	0

Next consider the power set $P(A)$ of a set S which is a singleton. Clearly $P(S)$ contains only two elements $\varnothing$ and S. Three tables similar to those shown above for the set-theoretic intersection, union and complementation of $\varnothing$ and S may be constructed as follows.

$\cap$	$\varnothing$	S
$\varnothing$	$\varnothing$	$\varnothing$
S	$\varnothing$	S

$\cup$	$\varnothing$	S
$\varnothing$	$\varnothing$	S
S	S	S

$\sim$	
$\varnothing$	S
S	$\varnothing$

The above two sets of tables of operations show that these two Boolean algebras are isomorphic to each other.

Example 1.7.3. Consider the four-element Boolean algebra $B_4 = \{0,a,b,1\}$. The algebra is described by the tables shown below.

·	0	a	b	1
0	0	0	0	0
a	0	a	0	a
b	0	0	b	b
1	0	a	b	1

+	0	a	b	1
0	0	a	b	1
a	a	a	1	1
b	b	1	b	1
1	1	1	1	1

	'
0	1
a	b
b	a
1	0

Now consider the power set $P(I)$, where $I = \{a,b\}$. The subsets of I are $\{a\}$, $\{b\}$, and the empty set. Denote them respectively by S_a, S_b, and $\varnothing$. As shown in Example 1.7.1, $P(I)$ is a Boolean algebra with the three operations shown in the following tables.

∩	$\varnothing$	S_a	S_b	I
$\varnothing$	$\varnothing$	$\varnothing$	$\varnothing$	$\varnothing$
S_a	$\varnothing$	S_a	$\varnothing$	S_a
S_b	$\varnothing$	$\varnothing$	S_b	S_b
I	$\varnothing$	S_a	S_b	I

∪	$\varnothing$	S_a	S_b	I
$\varnothing$	$\varnothing$	S_a	S_b	I
S_a	S_a	S_a	I	I
S_b	S_b	I	S_b	I
I	I	I	I	I

	~
$\varnothing$	I
S_a	S_b
S_b	S_a
I	$\varnothing$

Again, it is seen that B_4 is isomorphic to $P(I)$, where $I = \{a,b\}$.

From the above two examples, one may ask whether every Boolean algebra is isomorphic to a power-set algebra? The answer is yes. Before proceeding with the proof, we first need to introduce the following lemmas.

Definition 1.7.7. In a lattice with zero element 0, any element covering 0 is called an *atom*.

Let B be a Boolean algebra and x be an element of B. If we define $A(z)$ to be the set of all atoms a such that $a \leqslant x$, then it is clear that $A(0) = \varnothing$ and $A(1) = A$. Since B is a lattice, for any x other than 0, $A(x)$ is non-empty.

Lemma 1.7.1. Define A to be the set of all atoms of a Boolean algebra B. Then:

(a) $A(x) \cap A(x') = 0$

(b) $A(x) \cup A(x') = A$

(c) $A(x) = A(y)$ iff $x = y$ for all x,y in B

(d) $A(x) \subseteq A(y)$ iff $x \leq y$

Proof.

(a) *Proof of (a).* Suppose $A(x) \cap A(x') \neq 0$. Then there must exist an atom, say z, in the intersection; $z \in A(x)$ implies $z < x$, and $z \in A(x)$ implies $z \leq x'$. These jointly imply $z \cdot z = z \leq x \cdot x' = 0$ or $z = 0$. This is a contradiction; since z is an atom, $z \neq 0$. Hence $A(x) \cap A(x') = \varnothing$.

(b) *Proof of (b).* Let a be an element in A. Then $a \cdot x \leq a$. Since a is an atom, $a \cdot x$ is either a or 0. If $a \cdot x = a$, then $a \in A(x)$. If $a \cdot x = 0$, then $a \cdot x' = a$, since

$$a = a \cdot (x + x') = a \cdot x + a \cdot x' = a \cdot x'$$

hence $a \in A(x')$. Therefore $A(x) \cup A(x') = A$.

(c) *Proof of (c).* If $x = y$, it is clear that $A(x) = A(y)$. Conversely, assume that $A(x) = A(y)$. Let $x \neq y$. Then either $x \nleq y$, or $y \nleq x$, or both. If $x \nleq y$, then $x \cdot y \neq x$ which implies $x \cdot y' \neq 0$. This is because if $x \cdot y' = 0$, then

$$x = x \cdot y + xy' = x \cdot y$$

which contradicts the fact $x \cdot y \neq x$. If $x \cdot y' \neq 0$, then there must exist at least an atom a such that $a \leq x \cdot y'$, which in turn implies $a \leq x$ and $a \leq y'$. This means that $a \in A(x)$ and $a \in A(y')$. Since $a \in A(y')$, by (a) and (b), $a \notin A(y)$; then $A(x) \neq A(y)$ which contradicts our assumption. Thus $x \nleq y$ *cannot be the case. By symmetry,* $y \nleq x$ also implies $A(x) \neq A(y)$. Hence x must be equal to y.

(d) *Proof of (d).* If $x \leq y$, then for any atom a in $A(x)$
$$a \cdot x \leq x \leq y$$

Hence $A(x) \subseteq A(y)$. Conversely, assume that $A(x) \subseteq A(y)$, but $x \geq y$. Then, if a is an atom in $A(y)$, $a \leq y \leq x$, which means that a must also be in $A(x)$. This leads to the conclusion that $A(x) \supseteq A(y)$, which is a contradiction. Hence $x \leq y$.

Lemma 1.7.2. *For any x,y in a Boolean algebra,*

(a) $A(x \cdot y) = A(x) \cap A(y)$

(b) $A(x + y) = A(x) \cup A(y)$

(c) $A(x') = A - A(x)$

Proof. From the definition of $\cdot$, we have that $a \leq x \cdot y$ iff $a \leq x$ and $a \leq y$. If a is an atom in $A(x \cdot y)$, this relation implies that $a \in A(x)$

and $a \in A(y)$, or $A(x) \cap A(y)$; also if a is an element of $A(x) \cap A(y)$, the above relation implies that $a \in A(x \cdot y)$. The equation (b) may be shown using the relation $a \leqslant x + y$ iff $a \leqslant x$ or $a \leqslant y$. The equation (c) follows directly from Lemma 2.5.1(b).

Lemma 1.7.3. Let $a_1, a_2, \ldots, a_n$ be n distinct atoms of a Boolean algebra. Then

(a) $A(a_1 \cdot a_2 \cdot \ldots \cdot a_n) = A(a_1) \cap A(a_2) \cap \ldots \cap A(a_n) = \varnothing$

(b) $A(a_1 + a_2 + \ldots + a_n) = A(a_1) \cup A(a_2) \cup \ldots \cup A(a_n)$

$$= \{a_1, a_2, \ldots, a_n\}$$

Proof. Since the operations $\cdot$ and $+$ satisfy the associative laws, by repeatedly applying Lemma 1.7.2 (a) and (b) we obtain:

$$A(a_1 \cdot a_2 \cdot \ldots \cdot a_n) = A(a_1) \cap A(a_2) \cap \ldots \cap A(a_n)$$

and

$$A(a_1 + a_2 + \ldots + a_n) = A(a_1) \cup A(a_2) \cup \ldots \cup A(a_n)$$

Since the a_i are atoms, $A(a_i) = \{a_i\}$. This is because if in $A(a_i)$ there is another atom a_j, then $a_j \leqslant a_i$. Since we assume $a_j \neq a_i$, a_j must be less than a_i. In other words, $a_j = 0$. This is a contradiction. Hence

$$A(a_1 \cdot a_2 \cdot \ldots \cdot a_n) = \varnothing \qquad \text{and}$$

$$A(a_1 + a_2 + \ldots + a_n) = \{a_1, a_2, \ldots, a_n\}$$

From the above lemma, we have the following important theorems.

Theorem 1.7.2. Let $B = \{x_1, x_2, \ldots, x_n\}$ be a finite Boolean algebra. Define F to be the family of sets of $A(x_i)$, i.e.,

$$F = \{A(x_1), A(x_2), \ldots, A(x_n)\}$$

Then $(B, \cdot, +, ')$ and $(F, \cap, \cup, \sim)$ are isomorphic.

Proof. Since all x_i are distinct, from Lemma 1.7.1(c) the transformation $\phi = x_i \rightarrow A(x_i)$ from B to F is one to one and onto. Also, the transformation ϕ carries the Boolean operations $\cdot$, $+$, and $'$ into the set-theoretic operations $\cap$, $\cup$, and $\sim$, respectively, from B to F. Hence the theorem.

Theorem 1.7.3. Let A be the set of all atoms $a_1, a_2, \ldots, a_m$ of a Boolean algebra B. Then $F = P(A)$.

Proof. Obviously, every element of F is an element of $P(A)$. For the converse, consider an element in $P(A)$, say $A_k = a_1, a_2, \ldots, a_k$, where the a_i, are assumed to be distinct and $k \leqslant m$:

$$A_k = A(a_1) \cup A(a_2) \cup \ldots \cup A(a_k) = A(a_1 + a_2 + \ldots + a_k) = A(x_k)$$

which is in F, where $x_k = a_1 + a_2 + \ldots + a_k$. Hence $F = P(A)$.

Combining Theorems 1.7.2 and 1.7.3 we have the following theorem.

Theorem 1.7.4. Let A be the set of all atoms of a finite Boolean algebra B. Then B is isomorphic to $P(A)$.

Corollary 1.7.2. The number of elements of a finite Boolean algebra is a power of 2. In other words, for a given Boolean algebra B there exists a positive integer n such that the number of elements of B is 2^n.

Corollary 1.7.3. If B and C are two Boolean algebras with the same number of elements, then B and C are isomorphic.

Corollary 1.7.4. A finite 2^n-element Boolean algebra B has exactly n atoms.

Proof. By Theorem 1.7.4, B is isomorphic to $P(A)$, and it has been shown in Theorem 1.4.1 that, if A has n distinct elements, $P(A)$ has 2^n distinct elements.

Theorem 1.7.5. The Cartesian product of two sets B and C is a Boolean algebra if and only if L and M are Boolean algebras.

The proof is routine and is left to the reader.

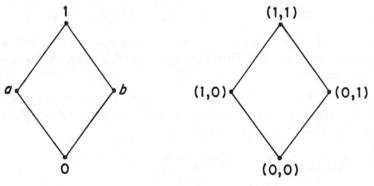

FIG. 1.7. Two Isomorphic lattices.

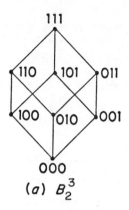

(a) B_2^3

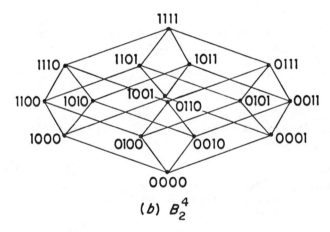

(b) B_2^4

FIG. 1.8. Two Cartesian product algebras.

Corollary 1.7.5. Let B and C be two Boolean algebras with finite m and n atoms, respectively. Then the Cartesian product algebra $B \times C$ has $2^{m \times n}$ elements.

Corollary 1.7.6. Any finite Boolean algebra B is isomorphic to a Cartesian product of two-element Boolean algebras $B_2 = \{0,1\}$, namely

$$B_2^n = \underbrace{B_2 \times B_2 \ldots B_2}_{n}$$

for some integer n.

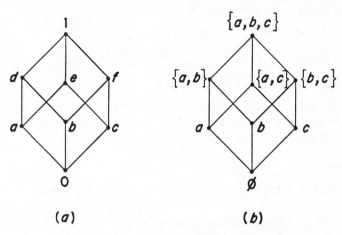

FIG. 1.9. An eight-element Boolean algebra.

For example, the four-element Boolean B_4 in Example 1.7.3 is isomorphic to

$$B_2^2 = B_2 \times B_2 = \{(0,0),(0,1),(1,0),(1,1)\}$$

as shown in Fig. 1.7. The Boolean algebras B_2^n, $n = 3$ and 4 are shown in Fig. 1.8.

Example 1.7.4. Consider the eight-element Boolean algebra shown in Fig. 1.9*a*. It is easy to find the products and sums of all the pairs of elements and the complement of each element of this algebra. The tables for the three operations are shown below.

·	0	a	b	c	d	e	f	1
0	0	0	0	0	0	0	0	0
a	0	a	0	0	a	a	0	a
b	0	0	b	0	b	0	b	b
c	0	0	0	c	0	c	c	c
d	0	a	b	0	d	a	b	d
e	0	a	0	c	a	e	c	e
f	0	0	b	c	b	c	f	f
1	0	a	b	c	d	e	f	1

+	0	a	b	c	d	e	f	1
0	0	a	b	c	d	e	f	1
a	a	a	d	e	d	e	1	a
b	b	d	b	f	d	1	f	b
c	c	e	f	c	1	e	f	c
d	d	d	d	1	d	1	1	d
e	e	e	1	e	1	e	1	e
f	f	1	f	f	1	1	f	f
1	1	a	b	c	d	e	f	1

′	
0	1
a	f
b	e
c	d
d	c
e	b
f	a
1	0

The atoms are a, b, and c, and the sets $A(x)$ are

$A(a) = \{a\}$ $A(d) = \{a,b\}$ $A(0) = \{\varnothing\}$

$A(b) = \{b\}$ $A(e) = \{a,c\}$ $A(1) = \{a,b,c\}$

$A(c) = \{c\}$ $A(f) = \{b,c\}$

The algebra is isomorphic to the power set $P(A)$, where $A = \{a,b,c\}$ as shown in Fig. 1.9b, as seen from the diagram and the relation between x and $A(x)$ shown above.

Exercises

(1) Show that a sublattice of a Boolean algebra need not be a sub-algebra.

(2) Show that the Cartesian product $P(A) \times P(B)$ of two power sets with finite m and n atoms is isomorphic to a power set with $m + n$ atoms.

(3) For any element x and y in a Boolean algebra:

$$(x + y) \cdot (x' + z) = x \cdot z + x' \cdot y$$

(4) In any Boolean algebra B:

  (a) $(x')' = x$

  (b) $(x \cdot y)' = x' + y'$

  (c) $(x + y)' = x' \cdot y'$

  (d) if $x \leqslant y$, then $y' \leqslant x'$

  (e) if $x' \leqslant y'$ then $y \leqslant x$

(5) Either give a Boolean algebra with three elements or show that none exists.

(6) If x is an element of the interval $[a,b]$ of a lattice L and $y \in L$ such that $x \cdot y = a$ and $x + y = b$, y is called a *complement of x relative to the interval* $[a,b]$. Show that a Boolean algebra is relatively complemented.

(7) A lattice for which every subset has both a g.l.b. and a l.u.b. is said to be *complete*. Prove that the infinite distributive laws

$$x \cdot (y_1 + y_2 + \ldots) = x \cdot y_1 + x \cdot y_2 + \ldots$$

$$x + (y_1 \cdot y_2 \cdot \ldots) = (x + y_1) \cdot (x + y_2) \cdot \ldots$$

hold in a complete Boolean algebra.

(8) Consider the eight-element Boolean algebra

$$B_8 = \{0,a,b,c,d,e,f,1\}$$

The multiplication, addition, and complementation tables are given below.

·	0	a	b	c	d	e	f	1
0	0	0	0	0	0	0	0	0
a	0	a	a	a	0	0	0	a
b	0	a	b	a	0	f	f	b
c	0	a	a	c	d	d	0	c
d	0	0	0	d	d	d	0	d
e	0	0	f	d	d	e	f	e
f	0	0	f	0	0	f	f	f
1	0	a	b	c	d	e	f	1

+	0	a	b	c	d	e	f	1
0	0	a	b	c	d	e	f	1
a	a	a	b	c	c	1	b	1
b	b	b	b	1	1	1	b	1
c	c	c	1	c	c	1	1	1
d	d	c	1	c	d	e	e	1
e	e	1	1	1	e	e	e	1
f	f	b	b	1	e	e	f	1
1	1	1	1	1	1	1	1	1

′	
0	1
a	e
b	d
c	f
d	b
e	a
f	c
1	0

(a) Find all the atoms of this algebra.

(b) Find the sets $A(0), A(a), \ldots, A(1)$.

(c) Draw the inclusion diagram of this algebra and show that it is isomorphic to a power-set algebra.

(9) Draw a picture of a lattice that is not a Boolean algebra. Prove that it is not by showing that it is not distributive or not complemented. For example, the following lattice *is* a Boolean algebra

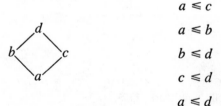

$$a \leq c$$
$$a \leq b$$
$$b \leq d$$
$$c \leq d$$
$$a \leq d$$

(10) Let ϕ be a transformation mapping from B_8 of Example 1.7.4 into itself which is defined by

$$\phi(0) = 0$$
$$\phi(a) = \phi(b) = \phi(c) = a$$
$$\phi(d) = \phi(e) = \phi(f) = c$$
$$\phi(1) = e$$

Show that ϕ is homomorphic; hence the image of ϕ is a Boolean algebra.

(11) A Boolean algebra B consists of eight elements $a, b, \ldots, h$. Its multiplication table is as shown below.

	a	b	c	d	e	f	g	h
a	a	g	a	a	g	g	g	a
b	g	b	b	b	g	b	g	g
c	a	b	c	c	g	b	g	a
d	a	b	c	d	e	f	g	h
e	g	g	g	e	e	e	g	e
f	g	b	b	f	e	f	g	e
g	g	g	g	g	g	g	g	g
h	a	g	a	h	e	e	g	h

The complementation table is:

$$a' = f, \quad b' = h, \quad c' = e, \quad d' = g$$

What is the addition table?

REFERENCES

Set Theory

Allendoerfer, C. B., and Oakley, C.O. *Principles of Mathematics.* McGraw-Hill, New York, 1955.

Birkhoff, G. The Lattice Theory. *American Mathematical Society Colloquium, New York, 1948*, vol. 25.

Breuer, J. *Introduction of the Theory of Sets.* Prentice-Hall, Englewood Cliffs, N.J., 1958.

Fraenkel, A. A. *Abstract Set Theory.* North-Holland, Amsterdam, 1953.

Halmos, P. R. *Naive Set Theory.* Van Nostrand, Princeton, N.J., 1960.

Kemeny, J. G., and Mirkil, H., Snell, J. L. and Thompson, G. L., *Finite Mathematical Structure.* Prentice-Hall, Englewood Cliffs, N.J. 1958.

Kleene, S. C. *Introduction to Metamathematics*. Van Nostrand, Princeton, N.J., 1952.

Mathematical Association of America, Committee on the Undergraduate Program. *Elementary Mathematics of Sets*. Ann Arbor, Mich. 1958.

Quine, W. V. O. *Set Theory and Its Logic*. Belknap Press of Harvard University Press, Cambridge, Mass., 1963.

Rotman, B., and Kneebone, G. T., *The Theory of Sets and Transfinite Numbers*. Spottiswoode, Ballentyne, London, 1966.

Rubin, J. E. *Set Theory for the Mathematician*. Holden-Day, San Francisco, 1967.

Rudin, W. *Principles of Mathematical Analysis*. McGraw-Hill, New York, 1953.

Taylor, A. E. *Introduction to Functional Analysis*. John Wiley, New York, 1958.

Zehna, P. W., and Johnson, R. L. *Elements of Set Theory*. Allyn and Bacon, Boston, Mass., 1962.

Lattice Theory

Abbott, J. C. *Sets, Lattices, and Boolean Algebras*. Allyn and Bacon, Boston, Mass., 1969.

Abbott, J. C. *Trends in Lattice Theory*. Van Nostrand, Princeton, N.J., 1969.

Birkhoff, G. *Lattice Theory*. (3rd edn). American Mathematical Society College Publications, Providence, R.I., 1967.

Donnellan, T. *Lattice Theory*. Pergamon Press, Oxford, 1968.

Dubisch, R. *Lattices to Logic*. Blaisdell, New York, 1963.

Fuchs, L. *Partially Ordered Algebraic Systems*. Pergamon Press, Oxford, 1963.

Lattice Theory, Proceedings of the Symposium on Pure Mathematics, vol. 2. American Mathematical Society, Providence, R.I., 1961.

Lieber, L. R. *Lattice Theory*. Galois Institute of Mathematics and Art, Brooklyn, N.Y., 1959.

MacLane, S., and Briknoff, G. *Algebra*, Chap. 14, pp. 482–505. Macmillan, New York, 1967.

Rutherford, D. E. *Introduction to Lattice Theory*. Hafner, New York, 1965.

Skornjakov, L. A. *Complemented Modular Lattices and Regular Rings*. Gos. Izd.-Mat. Lit., Moscow, 1961.

Szasz, G. *Introduction to Lattice Theory*. Academic Press, New York, and Akademiai Kiado, 1963.

Boolean Algebra

Abbott, J. C. *Sets, Lattices, and Boolean Algebras*. Allyn and Bacon, Boston, Mass. 1969.

Adelfio, S. A., Jr., and Nolan, C. F. *Principles and Applications of Boolean Algebra*. Hayden, New York, 1964.

Arnold, B. H. *Logic and Boolean Algebra*. Prentice-Hall, Englewood Cliffs, N.J., 1962.

Boole, G. *An Investigation into the Laws of Thought*. Open Court, Chicago, 1854/1940.

Bowran, A. P. *A Boolean Algebra*. Macmillan, London, 1965.

Flegg, H. G. *Boolean Algebra and its Applications*. John Wiley, New York, 1964.

Goodstein, R. L. *Boolean Algebra*. Pergamon Press, Oxford, 1966.

Halmos, P. R. *Lectures on Boolean Algebras*. Van Nostrand, Princeton, N.J., 1963.

Hohn, F. E. *Applied Boolean Algebra*. Macmillan, New York, 1960.

Sampath Kumarachar, E. *Some Studies in Boolean Algebra*. Karnatak University, Charwar, 1967.

Sikorski, R. *Boolean Algebras* Springer-Verlag, Berlin, 1964 (English edn).

Whitesitt, J. E. *Boolean Algebra and Its Applications*. Addison-Wesley, Reading, Mass., 1961.

CHAPTER 2

Fuzzy Sets, Logic, and Algebra

2.1. INTRODUCTION

In Chapter 1 we introduced *abstract* (or *conventional,* or *non-fuzzy*) set theory. In set theory the sets considered are abstract sets which are defined as collections of objects having some very general property *P;* nothing special is assumed or considered about the nature of the individual objects. For example, we define a set *A* as the set of cars. Symbolically: $A = \{x \,|\, x \text{ is a car}\}$.

Now what about the "class of *new* cars"? First of all, is it a set in the ordinary sense? Before we answer, we may first ask: "How 'new' is a new car? Is a one-year-old car a new car? If so, then is there any difference between a half-year-old new car and one-year-old new car?" Frankly, we do not know how to answer these questions adequately from the information *"new* cars," because the "class of new cars" does not constitute a set in the usual sense. Sets of this nature very often involve some adjectives, verbs, and adverbs or some combination thereof which are not sharply defined in their descriptions, such as the following.

(1) The class of *short* men. (Am I a member of this set?)

(2) The class of *new, high* buildings. (Is the United Nations Headquarters a member of this set?)

45

(3) The class of all real numbers which are *much* greater than 1. (Is 25 a member of this set?)
Numerous other such examples may be found in very nearly every branch of science and engineering, as well as in writings and daily conversations. In fact, most of the classes of objects encountered in the real physical world are of this *fuzzy, not sharply defined* type. They do not have precisely defined criteria of membership. In such classes an object need not necessarily either belong to or not belong to a class; there may be intermediate grades of membership. This is the concept of a fuzzy set, which is a "class" with a *continuum of grades of membership*.

This chapter is a collection of basic results of fuzzy set theory, algebra, and logic, written for the reader who finds this material completely or partially novel. Related and non-advanced topics are discussed in detail in many excellent papers listed in the bibliography.

2.2. DEFINITION OF A FUZZY SET

Fuzzy set theory, introduced by Zadeh, is a generalization of abstract set theory. In other words the former always includes the latter as a special case; definitions, theorems, proofs, etc., of fuzzy set theory always hold for non-fuzzy sets. Because of this generalization, fuzzy set theory has a much wider scope of applicability than abstract set theory in solving various kinds of *real physical world* problems, particularly in the fields of pattern classification, information processing, control, system identification, artificial intelligence, and, more generally, decision processes involving incomplete or uncertain data.

Following the concept of fuzziness introduced above, we now formally define a fuzzy set as follows.

Let X be a space of objects and x be a generic element of X. We shall call the set X the *object space*. Let $p_1, p_2, \ldots, p_n$ be n properties of x of interest which may be considered as n variables. If the n properties are *mutually unrelated*, they may be treated as n independent variables. We define the *property vector* as the n-tuple vector $(p_1, p_2, \ldots, p_n)$, and the *property space* denoted by P as the set of all possible values which the vector $(p_1, p_2, \ldots, p_n)$ can assume. Since we shall only be concerned with the n properties of x, to link the two spaces together we denote each point in the property space by $x = (p_1, p_2, \ldots, p_n)$.

Definition 2.2.1. A *fuzzy* set A in X is characterized by a *membership (characteristic) function* with respect to certain properties of x of interest, $p_1, p_2, \ldots, p_n$, denoted by

$$f_A(x = (p_1, p_2, \ldots, p_n)),$$

which is a functional mapping from the property space defined by the object space X into the interval $[0,1]$. The value of

$$f_A(x = (p_1, p_2, \ldots, p_n))$$

at x represents the *grade of membership* of x in A.

For simplicity, we shall sometimes write $f_A(x)$ instead of $f_A(x = (p_1, p_2, \ldots, p_n))$. However, it is assumed to be understood that when we are concerned with the n properties $p_1, p_2, \ldots, p_n$ of x in A, x will be considered as an n-tuple vector in P whose components are p_1, $p_2, \ldots, p_n$, and $f_A(x)$ is a function of variables $p_1, p_2, \ldots, p_n$.

From the above definition of a fuzzy set we observe that the nearer the value of $f_A(x)$ to unity, the higher the grade of membership of x in A. If A is a non-fuzzy set, then $f_A(x)$ is reduced to the familiar characteristic function A, which can take only two values, 0 and 1, with $f_A(x) = 1$ or 0 respectively as x does or does not belong to A. In fuzzy sets an element whose grade of membership is 1 will be said to have *full membership;* an element whose grade of membership is 0 will be said to have non-membership.

A fuzzy set may also be defined as a set of ordered pairs as follows.

Definition 2.2.2. Let X be a space of points (objects) and x be a generic element of X. Let $p_1, p_2, \ldots, p_n$ be n properties of x of interest. Then a *fuzzy* set A in X is a set of ordered pairs:

$$A = \{(x, f_A(x = (p_1, p_2, \ldots, p_n)) | x \in A \text{ and } f_A(x = (p_1, p_2, \ldots, p_n)) \text{ is the membership function described above}\}.$$

It should be noted that, although the membership function of a fuzzy set has some resemblance to a probability function when X is a countable set or a probability function when X is a continuum, they describe two different things—the grade of membership and the chance of having a certain outcome; thus their meanings are quite different. In fact, the notion of a fuzzy set is completely non-statistical in nature.

In the previous section several examples of "set" were given which we were unable to define as sets in the ordinary sense. Now if Definition 2.2.1 or Definition 2.2.2 is employed, they can all be defined as fuzzy sets.

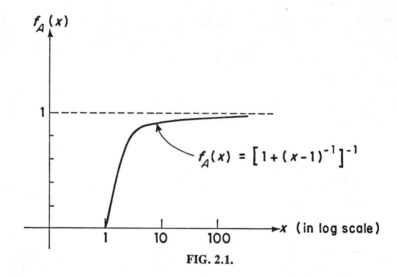

FIG. 2.1.

Example 2.2.1. The class of all real numbers which are much greater than 1.

We usually write this set as $A = \{x \mid x \text{ is a real number and } x \gg 1\}$. However, it is not a well-defined set for the reasons mentioned before. Such a set may be defined subjectively by a membership function such as

$$f_A(x) = 0 \qquad\qquad \text{for } x \leq 1$$
$$= \left[1 + (x - 1)^{-1}\right]^{-1} \text{for } x > 1$$

which is depicted in Fig. 2.1

It is interesting to notice that the universe of a fuzzy set could contain other fuzzy sets, and the "universes" of these other fuzzy sets might contain still other fuzzy sets, and so on. In this fashion one has higher "levels" of fuzzy sets, just as one has higher levels of ordinary sets. The level of a fuzzy set seems to measure the *abstractness* of a concept.

This observation permits us to model more sophisticated types of ambiguity and vagueness. For instance, the model of "short" which includes all fuzzy sets $f: X \rightarrow J$ defined by monotonically decreasing functions of height, can be given by the set $\overline{S}$ of all such J-sets. Better still, let $\overline{S}$ be the J-set of all J-sets defined by *approximately monotone decreasing functions; that is, the more nearly monotonic f, the greater the degree of membership of $f(h)$ in $\overline{S}$.* One might also wish to impose some second-derivative condition on functions in S, either fuzzy (i.e. approximate, as above) or sharp. In general, we prefer to remain flexible, allowing final choice to be tailored to the one particular situation at hand. For resolving the described paradox any one monotonic function, or the family of all of them, will do quite well.

2.3. OPERATIONS OF FUZZY SETS

In this section we begin with several definitions involving fuzzy sets which are obvious extensions of the corresponding definitions for ordinary sets given in Chapter 1.

Let X be a space of points (objects) and x be a generic element of X. Let $p_1, p_2, \ldots, p_n$ be n properties of x of interest.

Definition 2.3.1. A fuzzy set is *empty* iff its membership function is identically zero on X. A fuzzy set is *universal* iff its membership function is identically unity on X.

Example 2.3.1. Let X be the set of all real numbers greater than 1. Let A be the set of all real numbers less than 1. Then $f_A(x) = 0$ for all $x \in X$. Hence A is an empty set in X. On the other hand, if we let B be the set of all numbers greater than 0, then $f_B(x) = 1$ for all $x \in X$, and B is a universal set in X.

Definition 2.3.2. Two fuzzy sets A and B are *equal*, written as $A = B$, iff $f_A(x) = f_B(x)$ for all x in X.

For simplicity, we shall abbreviate the statement "$f_A(x) = f_B(x)$ for all x in X" by "$f_A = f_B$."

Definition 2.3.3. Let A and B be fuzzy sets. A is *contained* in B, written as $A \subseteq B$, if $f_A \leq f_B$. A is *strictly contained* in B, denoted by $A \subset B$, iff $f_A < f_B$. A is said to be a *subset* of B if $A \subseteq B$, and a *proper subset* of B if $A \subset B$.

Definition 2.3.4. The *absolute complement* of a fuzzy set A is denoted by $\overline{A}$ and is defined by

$$f_{\overline{A}} = 1 - f_A. \tag{2.3.1}$$

If A and B are fuzzy sets, the *relative complement* of A with respect to B, denoted by $B - A$, is defined by

$$f_{B-A} = f_B - f_A \tag{2.3.2}$$

provided that $f_B(x) \geq f_A(x)$.

Example 2.3.2. Let X be the set of all real numbers, and let A and B be the set of all real numbers which are much greater than 1 and the set of all integers which are much greater than 1, respectively. Suppose we assign the membership functions of A and B to be

$$f_A(x) = 0 \qquad\qquad\qquad \text{for } x \leq 1$$
$$= \left[1 + (x - 1)^{-1}\right]^{-1} \quad \text{for } x > 1 \tag{2.3.3}$$

and

$$f_B(x) = 0 \qquad\qquad\qquad \text{for } x \leq 1 \text{ or } x \text{ is not an integer}$$
$$= \left[1 + (x - 1)^{-1}\right]^{-1} \quad \text{for } x \text{ is an integer and } x > 1 \tag{2.3.4}$$

then the set B is contained in the set A, because $f_B \leq f_A$. If equation (2.3.4) were defined as

$$f_B(x) = 0 \qquad\qquad\qquad \text{for } x \leq 1, \text{ or } x \text{ is not an integer}$$
$$= 0.9\left[1 + (x - 1)^{-1}\right]^{-1} \text{ for } x \text{ is an integer and } x > 1 \tag{2.3.5}$$

then B is strictly contained in A.

The absolute complement of A is

$$f_{\bar{A}}(x) = 1 \qquad\qquad\qquad \text{for } x \leq 1$$
$$= 1 - \left[1 + (x - 1)^{-1}\right]^{-1} \text{ for } x > 1. \tag{2.3.6}$$

In plain words, the set $\bar{A}$ is the set of real numbers which are less than 1 or *not* much greater than 1. The absolute complement of the set B may be similarly defined.

Since $B \subseteq A$, the relative complement of B with respect to A is defined. From equations (2.3.3) and (2.3.4), we find that

$$f_{A-B}(x) \qquad = 0 \qquad\qquad\qquad \text{for } x \leq 1$$
$$= \left[1 + (x - 1)^{-1}\right]^{-1} \quad \text{for } x \text{ is } not \text{ an integer and } x > 1$$
$$= 0 \qquad\qquad\qquad \text{for } x \text{ is an integer and } x > 1.$$

Definition 2.3.5. The *union* of two fuzzy sets A and B with respective membership functions $f_A(x)$ and $f_B(x)$ is a fuzzy set C, written as $C = A \cup B$, whose membership function is related to those of A and B by

$$f_C(x) = \text{Max}\left[f_A(x), f_B(x)\right], \quad x \in X \tag{2.3.7}$$

or, in abbreviated form,

$$f_C = f_A + f_B.$$

Definition 2.3.6. The *intersection* of two fuzzy sets A and B with respective membership functions $f_A(x)$ and $f_B(x)$ is a fuzzy set C, written as $C = A \cap B$, whose membership function is related to those of A and B by

$$f_C(x) = \text{Min}\big[f_A(x), f_B(x)\big], \quad x \in X \tag{2.3.8}$$

or, in abbreviated form,

$$f_C = f_A \cdot f_B.$$

Example 2.3.3. Let X be the set of real numbers. Let A be the set of real numbers which are close to 1 and let the membership function of A be defined by

$$f_A(x) = \frac{1}{1 + (x - 1)^2}, \qquad x \in X. \tag{2.3.9}$$

Let B be the set of real numbers which are close to 2 and let the membership function of A be defined by

$$f_B(x) = \frac{1}{1 + (x - 2)^2}, \qquad x \in X. \tag{2.3.10}$$

The union of A and B is

$$f_{A \cup B}(x) = \text{Max}\big[f_A(x), f_B(x)\big]$$

$$= \frac{1}{1 + (x - 1)^2}, \qquad x \leq 1.5 \tag{2.3.11}$$

$$= \frac{1}{1 + (x - 2)^2}, \qquad x \geq 1.5.$$

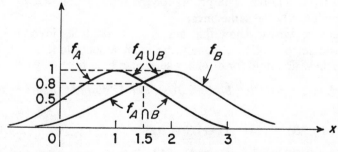

FIG. 2.2. Plots of f_A, f_B, $f_{A \cup B}$ and $f_{A \cap B}$ of equations (2.3.11) and (2.3.12).

since the two curves f_A and f_B intersect at $x = 1.5$ as shown in Fig. 2.2. Similarly, the intersection of A and B is

$$f_{A \cap B}(x) = \text{Min}\left[f_A(x), f_B(x)\right]$$

$$= \frac{1}{1 + (x - 2)^2}, \qquad x \leq 1.5 \qquad\qquad (2.3.12)$$

$$= \frac{1}{1 + (x - 1)^2}, \qquad x \geq 1.5.$$

Both $f_{A \cup B}$ and $f_{A \cap B}$ are shown in Fig. 2.2.

The union of these two sets means the set of all real numbers which are close to 1 *or* close to 2, and the intersection of these two sets means the set of all real numbers which are close to *both* 1 and 2.

Theorem 2.3.1. The union and intersection operations of two fuzzy sets are idempotent, commutative, and associative.

Proof. Let A, B, C be three fuzzy sets and f_A, f_B, f_C be the membership functions of A, B, and C resepctively. Since

$$f_A = \text{Max}(f_A, f_A)$$

and

$$\text{Max}(f_A, f_B) = \text{Max}(f_B, f_A)$$

$A = A \cup A$ and $A \cup B = B \cup A$. To prove that the union of two fuzzy sets is associate, we examine the following:

$$f_A + (f_B + f_C) = \text{Max}\left\{f_A, \text{Max}\left[f_B, f_C\right]\right\}$$

which is equal to

$$\text{Max}\left[\text{Max}(f_A, f_B), f_C\right] = (f_A + f_B) + f_C.$$

Hence it is associative. The proofs for the intersection of two fuzzy sets follow exactly the same lines.

Moreover, we can show that many basic identities involving the three basic set operations, namely the complementation, union, and intersection which hold for ordinary sets, also hold for fuzzy sets.

Theorem 2.3.2. (Distributive Laws). Let A, B, and C be fuzzy sets. Then

$$C \cap (A \cup B) = (C \cap A) \cup (C \cap B) \qquad\qquad (2.3.13a)$$

$$C \cup (A \cap B) = (C \cup A) \cap (C \cup B). \qquad\qquad (2.3.13b)$$

Proof. Let f_A, f_B, and f_C be the membership functions of A, B, and C, respectively. To prove equation (2.3.13a) is simply to prove that

$$\text{Min}\{f_C, \text{Max}[f_A, f_B]\} = \text{Max}\{\text{Min}[f_C, f_A], \text{Min}[f_C, f_B]\} \qquad (2.3.14)$$

which can be verified to be an identity by considering the six cases

(1) $f_A(x) \geqslant f_B(x) \geqslant f_C(x)$ (4) $f_B(x) \geqslant f_C(x) \geqslant f_A(x)$

(2) $f_A(x) \geqslant f_C(x) \geqslant f_B(x)$ (5) $f_C(x) \geqslant f_A(x) \geqslant f_B(x)$ (2.3.15)

(3) $f_B(x) \geqslant f_A(x) \geqslant f_C(x)$ (6) $f_C(x) \geqslant f_B(x) \geqslant f_A(x)$.

For example, for the first case, the left-hand side of equation (2.3.14) is

$$\text{Min}[f_C, \text{Max}(f_A, f_B)] = \text{Min}[f_C, f_A] = f_C$$

and the right-hand side is

$$\text{Max}[\text{Min}(f_C, f_A), \text{Min}(f_C, f_B)] = \text{Max}(f_C, f_C) = f_C.$$

The reader can easily verify equation (2.3.14) for the other five cases. Equation (2.2.13b) can be proven in a similar way.

Theorem 2.3.3. (De Morgan's Laws). Let A and B be two fuzzy sets. Then

$$(\overline{A \cup B}) = \overline{A} \cap \overline{B} \qquad (2.3.16)$$

$$(\overline{A \cap B}) = \overline{A} \cup \overline{B} \qquad (2.3.17)$$

Proof. To show equation (2.3.16) is to show that

$$1 - \text{Max}(f_A, f_B) = \text{Min}[(1 - f_A), (1 - f_B)] \qquad (2.3.18)$$

where f_A and f_B are the membership functions of A and B, respectively. If $f_A(x) > f_B(x)$, both sides of equation (2.3.18) are equal to $1 - f_A$. If $f_B(x) > f_A(x)$, both sides of equation (2.3.18) become $1 - f_B$. Hence equation (2.3.16) is an identity. Equation (2.3.17) may be proven similarly.

Besides the basic set operations introduced above, we can define the following additional set operations on fuzzy sets based on the algebraic operations of their membership functions.

Definition 2.3.7. The *symmetrical difference* (or *Boolean sum*) of two fuzzy sets A and B with membership functions f_A and f_B, denoted by $A \triangle B$, is a fuzzy set whose membership function $f_{A\triangle B}$ is related to those of A and B by

$$f_{A\triangle B} = |f_A - f_B|. \qquad (2.3.19)$$

Set operation	Symbol	Membership function
B is contained in A	$B \in A$	
The absolute complement of A	$\overline{A}$	
The relative complement of B with respect to A	$A - B$	
The union of A and B	$A \cup B$	
The intersection of A and B	$A \cap B$	
The symmetrical difference of A and B	$A \triangle B$	
The algebraic product of A and B	AB	

Set operation	Symbol	Membership function
The algebraic sum of A and B	$A + B$	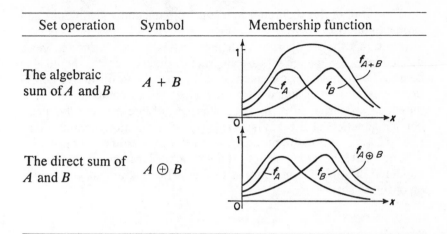
The direct sum of A and B	$A \oplus B$	

Let A and B be two fuzzy sets and $f_A(x)$ and $f_B(x)$ be the membership functions of A and B respectively. Here x represents both a given element of the space X and a specific property of x.

Definition 2.3.8. The *algebraic product* of two fuzzy sets whose membership function f_A and f_B, denoted by AB, is a fuzzy set whose membership function f_{AB} is related to those of A and B by

$$f_{AB} = f_A f_B. \qquad (2.3.20)$$

Definition 2.3.9. The *algebraic sum* of two fuzzy sets A and B with membership function f_A and f_B, denoted by $A + B$ is a fuzzy set whose membership function f_{A+B} is related to those of A and B by

$$f_{A+B} = f_A + f_B \qquad (2.3.21)$$

provided the sum $f_A(x) + f_B(x) \leqslant 1$ for all x.

Definition 2.3.10. The *direct sum* of two fuzzy sets A and B with membership functions f_A and f_B, denoted by $A + B$, is a fuzzy set whose membership function f_{A+B} is related to those of A and B by

$$f_{A \oplus B} = f_{A+B} - f_{AB} = f_A + f_B - f_A f_B. \qquad (2.3.22)$$

Theorem 2.3.4. Let A and B be fuzzy sets. Then

$$AB \subseteq A \cap B \subseteq A \cup B \subseteq A \oplus B \subseteq A + B. \qquad (2.3.23)$$

The proof of this theorem is evident from the above definitions and is thus omitted.

Note that the set operations defined in this section reduce to those defined in §1.2 when the sets are abstract (ordinary) sets. It should also be noted that the generic element x of the space X in the above definitions was always assumed to be a *vector* whose components are n unrelated properties of x, p_1, p_2, ..., and p_n, of interest, and the membership function of a fuzzy set A, $f_A(x)$, is a functional which maps $x = (p_1, p_2, \ldots, p_n)$ in the property space P into a real number in $[0,1]$.

For reference convenience, the set operations on fuzzy sets are tabulated in Table 2.1 in contrast to the Venn diagrams of operations on ordinary sets.

2.4. FUZZY RELATIONS

In Chapter 1 a relation on a set A was defined as a subset of $A \times A$. According to this definition, for any two members in A, they are either related or unrelated. For example, consider the relation "x is a son of y" among the people living in New York City. Using the above definition this relation is well defined, because if you choose any two persons in New York City they are either "father and son" or not. Now let us consider the relation "x is a close relative of y." Suppose we want to study this relation among the people living in New York City. Then we find that there are some difficulties in defining this relation. Unlike the previous case, choosing two persons in New York City we may face a situation where we are unable to decide whether they should be classified as related or unrelated; for example, they are brothers-in-law. This is mainly because, as mentioned before, the word "close" is not sharply defined.

Definition 2.4.1. A *fuzzy relation* L in X is a fuzzy set in $X \times X$, i.e.

$$L = \big((x,y), f_L(x,y)\big),\ x,y \in X,\ \text{and}\ 0 \leqslant f_L(x,y) \leqslant 1,$$

for all x and y in X. More generally, we can define an *n-ary fuzzy relation* in X as a fuzzy set L in the Cartesian product space $X \times X \ldots \times X$. For such relations the membership function is of the form $f_A(x_1, x_2, \ldots, x_n)$, where $x_i \in X$, $i = 1, 2, \ldots, n$.

Example 2.4.1. Let N be the set of natural numbers. Let $x, y \in N$.

The fuzzy relation "x is much greater than y" in N may be defined subjectively by a membership function such as

$$f_L(x,y) = \begin{cases} 0 & \text{if } x - y \leq 0 \\ [1 + 10(x - y)^{-1}]^{-1} & \text{if } x - y > 0. \end{cases}$$

Example 2.4.2. Let $x = (\alpha_1, \alpha_2, \ldots, \alpha_n)$ and $y = (\beta_1, \beta_2, \ldots, \beta_n)$ be two points in the n-dimensional Euclidean space R^n. The fuzzy relation "y is in the neighborhood of x" is a fuzzy set in R^n which may be defined subjectively by a membership function such as

$$f_L(x,y) = \frac{1}{\exp \|x - y\|}$$

where $\|x - y\| = [(\alpha_1 - \beta_1)^2 + (\alpha_2 - \beta_2)^2 + \ldots + (\alpha_n - \beta_n)^2]^{1/2}$.

Definition 2.4.2. The *composition* of two fuzzy relations A and B, denoted by $B \circ A$, is defined as a fuzzy relation in X whose membership function is related to those of A and B by

$$f_{B \circ A}(x,y) = \operatorname{Sup}_v \operatorname{Min}[f_A(x,v), f_B(v,y)], \quad v,x,y \in X.$$

Definition 2.4.3. Let X and Y be two spaces of objects, and h be a mapping from X to Y. Let B be a fuzzy set in Y with membership function $f_B(y)$. The fuzzy set A in X induced by the inverse mapping h^{-1} is defined by

$$A = \{(x, f_A(x)) \mid x = h^{-1}(y) \text{ and } f_A(x) = f_B(y), y \in B\}.$$

Now consider the converse problem. Suppose A is a given fuzzy set in X and h is a mapping from X to Y. What is the membership function for the fuzzy set B in Y which is induced by this mapping? To answer this question, we consider the following two cases separately.

Case A. The mapping h is one-to-one. The fuzzy set B in Y induced by the mapping h is defined by

$$B = \{(y, f_B(y)) \mid y = h(x) \text{ and } f_B(y) = f_A(x), x \in A\}.$$

Note that since h is one-to-one, for any $y \in Y$ where y is an image of some x in A, that is $y = h(x)$, there does not exist x' other than x such that $y = h(x')$. Hence $f_B(y)$ is uniquely defined by $f_A(x)$.

Case B. The mapping is many to one. In this case the following ambiguity arises. Suppose x_1 and x_2 are two distinct points with different grades of membership in A and are mapped to the same point y in Y.

Then what grade of membership in B should be assigned to y? To resolve this ambiguity we agree to assign the larger of the two grades of membership to y. If there are n distinct points $x_1, x_2, \ldots, x_n$ in A with different grades of membership which are mapped to the same point y in Y, the grade of membership in Y will be the largest of these grades of membership.

To combine the above two cases, we give the following general definition.

Definition 2.4.4. Let X and Y be two spaces of objects, and h be a mapping from X to Y. Let A be a fuzzy set in X with membership function $f_A(x)$. The fuzzy set B in Y induced by the mapping h is defined by

$$B = \{(y, f_B(y)) \mid y = h(x) \text{ and } f_B(y) = \underset{x \in T^{-1}(y)}{\text{Max}} f_A(x), x \in A\}$$

where $T^{-1}(y)$ is the set of points in X which are mapped into y by h.

2.5. CONVEX FUZZY SETS

In this section we shall be engaged in studying an important class of fuzzy sets, namely the convex fuzzy set. As will be seen in the sequel, the notion of convexity can readily be extended to fuzzy sets in such a way as to preserve many of the properties which it has in the context of ordinary sets discussed in §1.4. This notion appears to be particularly useful in applications involving pattern classification, optimization, and related problems.

We define a convex fuzzy set as follows.

Definition 2.5.1. Let X be a real linear space R^n. A fuzzy set A is *convex* if $x_1, x_2 \in X$, for all $\lambda \in [0,1]$,

$$f_A[\lambda x_1 + (1 - \lambda)x_2] \geq \text{Min}[f_A(x_1), f_A(x_2)]. \tag{2.5.1}$$

Example 2.5.1. Suppose a fuzzy set A whose membership function is defined by

$$f_A(x) = 0, \quad x < 0$$
$$= e^{-x}, \ x \geq 0$$

which is shown in Fig. 2.3. Without loss of generality, assume $x_2 \geq x_1$. Three possible cases are considered below.

(1) For $x_1, x_2 < 0$, trivially the condition of equation (2.5.1) is satisfied.

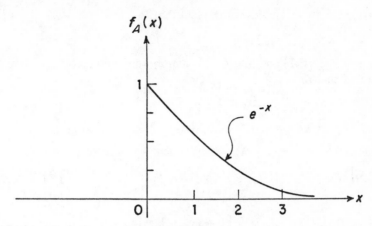

FIG. 2.3. The membership function of the fuzzy set A of Example 2.5.1.

(2) For $x_1 < 0$ but $x_2 > 0$ two subcases arise:

(2a) $x_1 \leqslant \lambda x_1 + (1 - \lambda)x_2 < 0$
and
(2b) $0 \leqslant \lambda x_1 + (1 - \lambda)x_2 \leqslant x_2$

since $x_1 \leqslant \lambda x_1 + (1 - \lambda)x_2 \leqslant x_2$. For the (2a) case equation (2.5.1) is satisfied; as for the (2b) case, the condition of equation (2.5.1) is also satisfied. Since

$$f_A[\lambda x_1 + (1 - \lambda)x_2] \geqslant \exp(-x_2) > 0 = \text{Min}[f_A(x_1), f_A(x_2)].$$

(3) For $x_1, x_2 \geqslant 0$

$$f_A[\lambda x_1 = (1 - \lambda)x_2] = \exp\{-[\lambda x_1 + (1 - \lambda)x_2]\}$$
$$\geqslant \exp\{-[\lambda x_2 + (1 - \lambda)x_2]\}$$
$$= \exp(-x_2) = \text{Min}[f_A(x_1), f_A(x_2)]$$

since $f_A(x_1) \geqslant f_A(x_2)$. Hence A is a convex fuzzy set.

Example 2.5.2. Consider the fuzzy set of Example 2.2.1 whose membership is defined by

$$f_A(x) = 0 \qquad \qquad \text{for } x \leqslant 1$$
$$= [1 + (x - 1)^{-1}]^{-1} \quad \text{for } x > 1.$$

This is a monotonically increasing function; that is, if $x_2 \geqslant x_1$,

$$f_A(x_2) \geqslant f_A(x_1).$$

Assume $x_2 \geq x_1$. In view of the fact that

$$x_1 \leq \lambda x_1 + (1 - \lambda)x_2 \leq x_2$$

for any $\lambda \in [0,1]$ and that $f_A(x)$ is a monotonically increasing function, we find that for

$$x_1 \leq \lambda x_1 + (1 - \lambda)x_2 \leq 1,$$

$$f_A[\lambda x_1 + (1 - \lambda)x_2] = 0 \geq f_A(x_1) = \text{Min}[f_A(x_1), f_A(x_2)],$$

and for $1 < \lambda x_1 + (1 - \lambda)x_2 \leq x_2$,

$$\begin{aligned}
f_A[\lambda x_1 + (1 - \lambda)x_2] &= \{1 + [(\lambda x_1 + (1 - \lambda)x_2) - 1]^{-1}\}^{-1} \\
&\geq \{1 + [(\lambda x_1 + (1 - \lambda)x_1) - 1]^{-1}\}^{-1} \\
&= [1 + (x_1 - 1)^{-1}]^{-1} = \text{Min}[f_A(x_1), f_A(x_2)].
\end{aligned}$$

Thus A is convex.

From the above examples we see that the necessary and sufficient condition for a fuzzy set to be convex is that its membership function does not have local minima, or the plot of its membership function does not have a section of concave curve.

A glance at the set operations tabulated in Table 2.1 indicates that the only set operation which preserves the convexity property is the intersection.

Theorem 2.5.1. Let A and B be two convex fuzzy sets in X. Then the intersection of A and B is convex.

Proof. Let f_A and f_B be the membership functions of A and B respectively. By hypothesis

$$f_A(\lambda x_1 + (1 - \lambda)x_2) \geq \text{Min}[f_A(x_1), f_A(x_2)]$$

$$f_B(\lambda x_1 + (1 - \lambda)x_2) \geq \text{Min}[f_B(x_1), f_B(x_2)]$$

and the membership function of the intersection C of A and B evaluated at $x = \lambda x_1 + (1 - \lambda)x_2$ is

$$\begin{aligned}
f_C(\lambda x_1 + (1 - \lambda)x_2) &= \text{Min}\{f_A[\lambda x_1 + (1 - \lambda)x_2], f_B[\lambda x_1 + (1 - \lambda)x_2]\} \\
&\geq \text{Min}\{\text{Min}[f_A(x_1), f_A(x_2)], \text{Min}[f_B(x_1), f_B(x_2)]\} \\
&\geq \text{Min}\{\text{Min}[f_A(x_1), f_B(x_1)], \text{Min}[f_A(x_2), f_B(x_2)]\} \\
&= \text{Min}[f_C(x_1), f_C(x_2)].
\end{aligned}$$

Thus the intersection of two convex fuzzy sets is convex.

Now, if we impose the condition that one convex set is contained in the other on the two convex sets A and B in Theorem 2.5.1, besides the intersection the only other operation that preserves the convexity property is the union.

The convex combination of ordinary sets was defined in §1.4 (Definition 1.4.5) which can be generalized to fuzzy sets in the following manner.

Definition 2.5.2. Let A, B, and Λ be fuzzy sets in X. The convex combination of A, B, and Λ is denoted by $(A,B;\Lambda)$ and is defined by the relation

$$(A,B;\Lambda) = A + \overline{\Lambda}B \qquad (2.5.2)$$

where $\overline{\Lambda}$ is the complement of Λ. Expressing equation (2.5.2) in terms of membership functions, we have

$$f_{(A,B;\Lambda)}(x) = f_\Lambda(x)f_A(x) + \left[1 - f_\Lambda(x)\right]f_B(x), \quad x \in X. \qquad (2.5.3)$$

A basic property of convex combination of fuzzy sets is given in the following theorem.

Theorem 2.5.2. $A \cap B \subseteq (A,B;\Lambda) \subseteq A \cup B$ for all Λ.

Proof. Assume $f_B \leqslant f_A$ and $\lambda \in [0,1]$. The following inequality holds.

$$f_B \leqslant f_B + \lambda(f_A - f_B) = \lambda f_A + (1 - \lambda)f_B \leqslant \lambda f_A + (1 - \lambda)f_A = f_A,$$

that is

$$f_B \leqslant \lambda f_A + (1 - \lambda)f_B \leqslant f_A. \qquad (2.5.4)$$

If $f_A \leqslant f_B$,

$$f_A = \lambda f_A + (1 - \lambda)f_A \leqslant f_A + (1 - \lambda)f_B = f_B - \lambda(f_B - f_A) \leqslant f_B,$$

that is

$$f_A \leqslant f_A + (1 - \lambda)f_B \leqslant f_B. \qquad (2.5.5)$$

Combining equations (2.5.4) and (2.5.5) we conclude that

$$\text{Min}\left[f_A(x),f_B(x)\right] \leqslant \lambda f_A(x) + (1 - \lambda)f_B(x) \leqslant \text{Max}\left[f_A(x),f_B(x)\right] \quad (2.5.6)$$

for all $x \in X$ and $\lambda \in [0,1]$. Hence

$$A \cap B \subseteq (A,B;\Lambda) \subseteq A \cup B.$$

It is worth noting that given a fuzzy set C satisfying

$$A \cap B \subseteq C \subseteq A \cup B,$$

one can always find a fuzzy set Λ such that $C = (A,B;\Lambda)$. In accordance with equation (2.5.5) the membership function $f_\Lambda(x)$ of this set is given by

$$f_\Lambda(x) = \frac{f_C(x) - f_B(x)}{f_A(x) - f_B(x)}, \qquad x \in X. \tag{2.5.7}$$

Example 2.5.3. Let A, B be two fuzzy sets in X whose membership functions are given by equations (2.3.9) and (2.3.10) respectively. The membership functions of the union and intersection of the two sets were given by equations (2.3.11) and (2.3.12). Suppose a fuzzy set C whose membership function $f_C(x)$ is defined by

$$f_C(x) = \begin{cases} \dfrac{1}{1 + (x - 1)^2} - \epsilon, & x \leq 1.5 \\[3mm] \dfrac{1}{1 + (x - 2)^2} - \epsilon, & x \geq 1.5 \end{cases} \tag{2.5.8}$$

where

$$0 \leq \epsilon \leq \left| \frac{1}{1 + (x - 1)^2} - \frac{1}{1 + (x - 2)^2} \right|.$$

Since the set C satisfies the relation $A \cap B \subseteq C \subseteq A \cup B$, it may be expressed as a convex combination of A and B. From equation (2.5.8) we find that the membership function of the corresponding fuzzy set Λ is

$$f_\Lambda(x) = \begin{cases} 1 - \dfrac{\epsilon[1 + (x - 1)^2][1 + (x - 2)^2]}{(x - 2)^2 - (x - 1)^2}, & x \leq 1.5 \\[3mm] \dfrac{\epsilon[1 + (x - 1)^2][1 + (x - 2)^2]}{(x - 1)^2 - (x - 2)^2}, & x \geq 1.5. \end{cases} \tag{2.5.9}$$

The definition of convex combination of two fuzzy sets may be extended to n fuzzy sets as in the ordinary set case.

Definition 2.5.3. Let $A_1, A_2, \ldots, A_n$ and $\Lambda_1, \Lambda_2, \ldots, \Lambda_n$ be fuzzy sets of X. The *convex combination* of $A_1, A_2, \ldots, A_n$ denoted by $(A_1, A_2, \ldots, A_n; \Lambda_1, \Lambda_2, \ldots, \Lambda_n)$ is a fuzzy set

$$(A_1, A_2, \ldots, A_n; \Lambda_1, \Lambda_2, \ldots, \Lambda_n) = \sum_{i=1}^{n} \Lambda_i A_i \qquad (2.5.10)$$

where the membership functions $f_{\Lambda_i}(x)$ of the fuzzy sets Λ_i satisfy the condition

$$\sum_{i=1}^{n} f_{\Lambda_i}(x) = 1.$$

For convenience, we shall abbreviate $(A_1, A_2, \ldots, A_n; \Lambda_1, \Lambda_2, \ldots, \Lambda_n)$ by $(A_i; \Lambda_i)_n$.

The property of the convex combination of two fuzzy sets presented in Theorem 2.5.2 is readily extended to the following more general theorem.

Theorem 2.5.3.

$$\bigcap_{i=1}^{n} A_i \subseteq (A_i; \Lambda_i)_n \subseteq \bigcup_{i=1}^{n} A_i$$

Proof. Let $f_{A_i}(x)$ and $f_{\Lambda_i}(x)$ be the membership functions of A_i and Λ_i. Suppose $f_{A_j}(x_1)$ and $f_{A_k}(x_1)$ are, respectively, the smallest and the largest among all $f_{A_i}(x_1)$, at $x = x_1$. We find that

$$f_{A_j}(x_1) = \sum_{i=1}^{n} \lambda_i f_{A_i}(x_1) - \left[\sum_{i=1}^{n} \lambda_i f_{A_i}(x_1) - f_{A_j}(x_1) \right]$$

$$\leq \sum_{i=1}^{n} \lambda_i f_{A_i}(x_1) - \left[\left(\sum_{i=1}^{n} \lambda_i \right) f_{A_j}(x_1) - f_{A_j}(x_1) \right]$$

$$= \sum_{i=1}^{n} \lambda_i f_{A_i}(x_1)$$

$$= f_{A_k}(x_1) - \left[f_{A_k}(x_1) - \sum_{i=1}^{n} \lambda_i f_{A_i}(x_1) \right]$$

$$\leq f_{A_k}(x_1) - \left[f_{A_k}(x_1) - \left(\sum_{i=1}^{n} \lambda_i \right) f_{A_k}(x_1) \right]$$

$$= f_{A_k}(x_1)$$

or

$$f_{A_j}(x_1) \leq \sum_{i} \lambda_i f_{A_i}(x_1) \leq f_{A_k}(x_1) \qquad (2.5.11)$$

where $0 \leqslant \lambda_i \leqslant 1$ and $\sum_{i=1}^{n} \lambda_i = 1$. Since x_1 is just an arbitrary x, from the inequality (2.5.11) we conclude that

$$\bigcap_{i=1}^{n} A_i \subseteq (A_i;\Lambda_i)_n \subseteq \bigcup_{i=1}^{n} A_i. \qquad (2.5.12)$$

Hence the theorem is proved.

Again, any fuzzy set C in X with membership function f_C satisfies

$$\bigcap_{i=1}^{n} A_i \subseteq C \subseteq \bigcup_{i=1}^{n} A_i;$$

it is always expressible as a convex combination of A_i. The membership functions of f_{Λ_i} and f_{A_i} are related by

$$\sum_{i=1}^{n} f_{\Lambda_i} f_{A_i} = f_C \qquad (2.5.13)$$

where $\sum_{i=1}^{n} f_{\Lambda_i} = 1$. It is clear that if, given m fuzzy sets C_j, we are asked to find $f_{\Lambda_{ij}}$ such that

$$C_j = (A_i; f_{\Lambda_{ij}})_n, \qquad j = 1, 2, \ldots, m \qquad (2.5.14)$$

there are three possible cases: (1) if $m < n - 1$, there are infinitely many solutions; (2) if $m = n - 1$, there is a unique solution; (3) if $m > n - 1$, there is no solution.

Definition 2.5.4. A fuzzy set is *bounded* if and only if the sets

$$S_\alpha = \left\{ x \,\middle|\, f_A(x) \geqslant \alpha \right\}$$

are bounded for all $\alpha > 0$. In other words, for every $\alpha > 0$ there exists a finite number $K(\alpha)$ such that the norm $\|x\|$ of every element x of S_α is less than or equal to $K(\alpha)$.

Theorem 2.5.4. Let A be a bounded fuzzy set and M be the maximal grade in A, i.e. $M = \mathrm{Sup}_x f_A(x)$. Assume $M > 0$. Then there exists at least one point x_0 at which M is essentially attained in the sense that, for each $\epsilon > 0$, every spherical neighborhood of x_0 contains points in the set

$$Q(\epsilon) = \left\{ x \,\middle|\, f_A(x) \geqslant M - \epsilon \right\}.$$

Definition 2.5.5. Let A be a bounded fuzzy set in X. The set of all points in X at which M is essentially attained is called the core of A, denoted by $C(A)$.

Using Theorem 2.5.4 it is easy to prove the following result.

Theorem 2.5.5. Let A be a convex fuzzy set. Then $C(A)$ *is convex.*

2.6. PROBABILITY MEASURES OF FUZZY EVENTS

The most basic concepts of probability theory originate from the notions of an event and its probability. An event A in probability theory is a subset of a sample space S. A probability measure P on S is a normed mapping from subsets A of S to real numbers such that

P1. For every $A \in S$, $P(A) \geqslant 0$, $P(\emptyset) = 0$, $P(S) = 1$.
P2. If A_1, $A_2 \in S$ and $A_1 \cap A_2 = \emptyset$, then

$$P(A_1 > A_2) = P(A_1) + P(A_2).$$

P3. If $\{A_1, A_2, \ldots,\}$ is a countable collection of disjoint subsets of S, then

$$P \bigcup_{i=1}^{\infty} A_i = \sum_{i=1}^{\infty} P(A_i).$$

As defined above an event is a precisely specified collection of points in the sample space. By contrast, in everyday experience one frequently encounters situations in which an "event" is a fuzzy rather than a sharply defined collection of points, for example, finding the probabilities of events such as the following.

(1) An ordinary die is thrown. What is the probability that the number which turns up is *close to 3?*

(2) In 100 tosses of a coin, what is the probability of having several more *heads than tails?*

(3) Suppose the weather of a city x has the transition probability matrix

$$P = [p_{ij}] = \begin{array}{c} \\ R \\ S \\ C \\ SN \end{array} \overset{\displaystyle R \quad S \quad C \quad SN}{\begin{bmatrix} \frac{1}{2} & \frac{1}{4} & \frac{1}{8} & \frac{1}{8} \\ \frac{1}{8} & \frac{5}{8} & \frac{1}{8} & \frac{1}{8} \\ \frac{3}{16} & \frac{3}{16} & \frac{7}{16} & \frac{3}{16} \\ \frac{1}{8} & \frac{1}{8} & \frac{1}{4} & \frac{1}{2} \end{bmatrix}} \qquad (2.6.1)$$

where R, S, C, and SN denote a rainy, sunny, cloudy, and snowy day. The entries in the first row represent the probabilities for the various kinds of weather following a rainy day, and those in the second, third, and fourth rows represent these probabilities following a sunny, cloudy, and snowy day, respectively. Suppose today's weather of City X is rainy. What is the probability of having *good weather* two days from today?

In the above examples "close to 3," "several more heads than tails," and "good weather" are *fuzzy* events. By using the concept of a fuzzy set, the notions of an event and its probability can be extended in a natural fashion to fuzzy events. It is possible that such an extension may eventually significantly enlarge the domain of applicability of probability theory, especially in those fields in which fuzziness is a pervasive phenomenon. Formally we define the following.

Definition 2.6.1. Let S be a real linear space R^n. A *fuzzy event A* in R^n is a fuzzy set in R^n.

Now we are ready to define the probability measure of a fuzzy event.

Definition 2.6.2. The *probability* of a fuzzy event A in R^n with membership function $f_A(x)$ is defined by

$$P(A) = \int_{R^n} f_A(x)\, dP$$

$$= E(f_A). \tag{2.6.2}$$

In other words, the probability of a fuzzy event is the expectation $E(f_A)$ of its membership function.

Following this definition, we immediately have the following theorem.

Theorem 2.6.1. Let A, B be two fuzzy events in R^n. Then

$(a)\ P(A \cup B) = P(A) + P(B) - P(A \cap B) \tag{2.6.3}$

$(b)\ P(A \oplus B) = P(A) + P(B) - P(AB). \tag{2.6.4}$

Proof. By Definitions 2.3.5 and 2.3.6,

$$f_{A \cup B}(x) = \text{Max}\left[f_A(x), f_B(x)\right]$$
$$f_{A \cap B}(x) = \text{Min}\left[f_A(x), f_B(x)\right]$$

thus

$$f_{A \cup B}(x) + f_{A \cap B}(x) = f_A(x) + f_B(x)$$
$$f_{A \cup B}(x) = f_A(x) + f_B(x) - f_{A \cap B}(x).$$

By Definition 2.6.2

$$P(A \cup B) = \int_{R^n} f_{A \cup B}(x) \, dP$$

$$= \int_{R^n} \left[f_A(x) + f_B(x) - f_{A \cap B}(x) \right] dP$$

$$= \int_{R^n} f_A(x) \, dP + \int_{R^n} f_B(x) \, dP - \int_{R^n} f_{A \cap B}(x) \, dP$$

$$= P(A) + P(B) - P(A \cap B).$$

Hence (a) is proved. The proof of (b) follows directly from the definition of $\oplus$ given in Definitions 2.3.10 and 2.6.2.

Now we want to define some of the quantities in probability theory for fuzzy events. Among them the one which establishes a basic concept in probability theory is the statistical independence. First, let us review the definition and meaning of statistical independence in probability theory of non-fuzzy events. It is known that probability theory of non-fuzzy events is based on ordinary set theory in which an element either "belongs" to a set or "does not belong to" it. Thus $A \cap B = \varnothing$ means that A and B are disjoint. If A and B are two events in a sample space, then the meaning of disjoint becomes "unrelated" or "independent." The probability theory of fuzzy events, on the other hand, depends on fuzzy set theory in which an element need neither "belong to" nor "not belong to." Consequently, if A,B are two fuzzy events, $A \cap B = \varnothing$ implies $f_A(x) = 0$, or $f_B(x) = 0$, or both $f_A(x) = 0$ and $f_B(x) = 0$ for every x. Then $P(A \cap B) = P(A)P(B)$ implies both sides of the equation are identically equal to zero, which does not express the meaning of independence we want. The meaning of two events A and B being statistically independent in probability theory is that the occurrence of an event A does not depend on the occurence of B and *vice versa*. Since in fuzzy set theory there is no such concept corresponding to "belonging to" and "not belonging to" in a sharp way, we therefore must define it differently.

Now consider a probability measure P defined on a sample space S. P is a Cartesian product of two measures P_1 and P_2 which are respectively defined on S_1 and S_2, and $S = S_1 \times S_2$. Let A_1 and A_2 be two fuzzy events defined on S_1 and S_2 with membership functions

$$f_{A_1}(x_1, x_2) = f_{A_1}(x_1)$$

$$f_{A_2}(x_1, x_2) = f_{A_2}(x_2).$$

Observe that the occurrences of events A and B are "unrelated." Moreover, A and B satisfy the following relation:

$$P(AB) = \int_S f_{A_1}(x_1,x_2)f_{A_2}(x_1,x_2)\, d(P_1 \times P_2)$$

$$= \int_{S_1 \times S_2} f_{A_1}(x_1)f_A(x_2)(dP_1 \times dP_2)$$

$$= \int_{S_1} f_{A_1}(x_1)\, dP_1 \cdot \int_{S_2} f_{A_2}(x_2)\, dP_2 = P(A)P(B).$$

Thus we define the following.

Definition 2.6.3. Two fuzzy events A and B are said to be statistically independent if

$$P(AB) = P(A)P(B). \tag{2.6.5}$$

This definition enables us to make a further definition.

Definition 2.6.4. Let A and B be two fuzzy events in a sample space R^n. The *conditional probability* of A given B is defined by

$$P(A \mid B) = \frac{P(AB)}{P(B)}. \tag{2.6.6}$$

Note that if A and B are statistically independent, then

$$P(A \mid B) = P(A)$$
and
$$P(B \mid A) = P(B).$$

Furthermore, we define the following.

Definition 2.6.5. Let A be a fuzzy event in a sample space R^1. The *mean* (or expected value) $m(A)$ of fuzzy event A is

$$m(A) = \int_{R^1} x f_A(x)\, dP = E(x f_A) \tag{2.6.7}$$

and the *variance* of A is

$$v(A) = \int_{R^1} [x - m(A)]^2 f_A(x)\, dP = E\{[x - m(A)]^2 f_A(x)\}. \tag{2.6.8}$$

Example 2.6.1. The sample space S of throwing a fair die is $S = \{1, 2, 3, 4, 5, 6\}$. The probability $p(x_i)$ for each of the outcomes is $\frac{1}{6}$, where $x_i = i$, $i = 1, 2, \ldots, 6$. Let A be a fuzzy event.

A is the number that the die turns up which is close to 3. Suppose A is described by a membership function $f_A(x)$: $f_A(1) = 0.4$, $f_A(2) = 0.6$, $f_A(3) = 1$, $f_A(4) = 0.6$, $f_A(5) = 0.4$, and $f_A(6) = 0.2$. Then the probability that the number which turns up is close to 3 is

$$P(A) = \sum_{i=1}^{6} f_A(x_i) \cdot p(x_i) = \tfrac{1}{6} \times 0.4 + \tfrac{1}{6} \times 0.6 + \tfrac{1}{6} \times 1 + \tfrac{1}{6} \times 0.6$$
$$+ \tfrac{1}{6} \times 0.4 + \tfrac{1}{6} \times 0.2$$
$$= 0.533.$$

The mean or expected value of A is calculated by

$$m(A) = \sum_{i=1}^{6} x_i f_A(x_i) p(x_i)$$
$$= \tfrac{1}{6}(1 \times 0.4 + 2 \times 0.6 + 3 \times 1 + 4 \times 0.6 + 5 \times 0.4 + 6 \times 0.2)$$
$$= 1.7.$$

The *variance* of A is

$$v(A) = \sum_{i=1}^{6} \left[x_i - m(A) \right]^2 f_A(x_i) p(x_i)$$
$$= \tfrac{1}{6} \big[0.49 \times 0.4 + 0.09 \times 0.6 + 1.69 \times 1 + 5.29 \times 0.6$$
$$+ 10.89 \times 0.4 + 18.49 \times 0.2 \big]$$
$$= 2.191.$$

Example 2.6.2. Consider the weather of the city x. The transition probability matrix was given in equation (2.6.1). The sample space S of this example is clearly $\{R,S,C,SN\}$. Suppose a fuzzy event A = good weather is defined in S whose membership function is subjectively defined by

$$f_A(R) = 0.2, \, f_A(S) = 1, \, f_A(C) = 0.5, \text{ and } f_A(SN) = 0.1.$$

As mentioned before the meaning of the transition probability matrix, for example the first row, is the probabilities for various kinds of weather following a rainy day. The probabilities for the various kinds of weather two days after a rainy day may be calculated by

$$\begin{bmatrix} \tfrac{1}{2} & \tfrac{1}{4} & \tfrac{1}{8} & \tfrac{1}{8} \end{bmatrix} \begin{bmatrix} \tfrac{1}{2} & \tfrac{1}{4} & \tfrac{1}{8} & \tfrac{1}{8} \\ \tfrac{1}{8} & \tfrac{5}{8} & \tfrac{1}{8} & \tfrac{1}{8} \\ \tfrac{3}{16} & \tfrac{3}{16} & \tfrac{7}{16} & \tfrac{3}{16} \\ \tfrac{1}{8} & \tfrac{1}{4} & \tfrac{1}{4} & \tfrac{1}{2} \end{bmatrix} = \begin{bmatrix} \tfrac{41}{128} & \tfrac{41}{128} & \tfrac{23}{128} & \tfrac{23}{128} \end{bmatrix}.$$

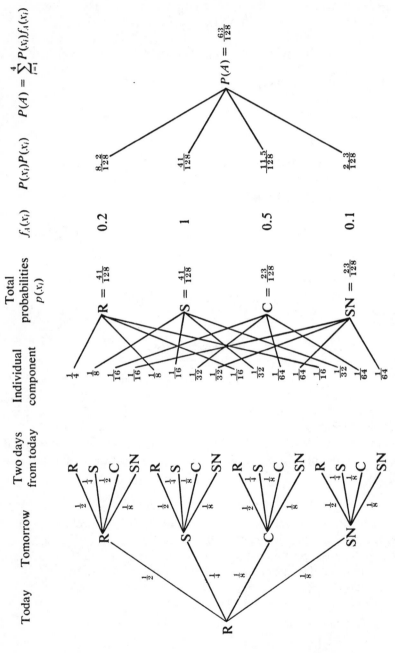

FIG. 2.4. A tree diagram to illustrate the procedure for calculating $P(A)$.

Therefore the probability of having good weather two days from today is

$$P(A) = \sum_{i=1}^{4} f_A(x_i)p(x_i)$$

$$= \tfrac{1}{128}(41{\times}0.2 + 41{\times}1 + 23{\times}0.5 + 23{\times}0.1) = \tfrac{63}{128}.$$

The entire procedure may be represented by a tree diagram which is shown in Fig. 2.4.

2.7 FUZZY LOGIC

The strictly binary approach to the treatment of switching theory today is not always adequate to describe systems in the real world. This approach is partly due firstly to the relative simplicity of designing binary switching systems, and secondly to the fact that basic switching modules in common use are two positional. Consequently, every variable in Boolean logic is assumed to be two valued. However, because of real-world constraints, the attributes of system variables are often ambiguously defined. In other words, quite often variables might have values other than falsehood and truth. Cases with such attributes arise, for example, in artificial intelligence and related subjects. It is our aim in this section to develop the basic fuzzy logic as a means for such investigations. Fuzzy logic is based on the concepts of fuzzy sets and symbolic logic.

We may view fuzzy logic as a special kind of many-valued logic. In fuzzy logic the truth value of a formula, instead of assuming two values (0 and 1), can assume any value in the interval $[0,1]$ and is used to indicate the degree of truth represented by the formula. For example, let $P(x)$ represent "x is a large number compared with unity"; then the truth value of $P(10^6)$ and $P(10^{-6})$ are certainly 1 and 0 respectively. As for $P(125)$, the truth value of it may be some value between 0 and 1, say 0.25.

We shall assume that well-formed formulas (formulas for short) are defined to be exactly the same as those in two-valued logic. Letting $T(S)$ denote the truth value of a formula s, the evaluation procedure for a formula in fuzzy logic can be described as follows.

(1) $T(S) = T(A)$ if $S = A$ and A is a ground atomic formula.
(2) $T(S) = 1 - T(R)$ if $S = \overline{R}$.
(3) $T(S) = \mathrm{Min}[T(S_1), T(S_2)]$ if $S = S_1 \cdot S_2$.

(4) $T(S) = \text{Max}[T(S_1). \ T(S_2)]$ if $S = S_1 + S_2$.

(5) $T(S) = \inf[T(B(x)) \times D]$ if $S = (x) \ B$ and D is the domain of x.

(6) $T(S) = \sup[T(B(x)) \times D]$ if $S = (Ex) \ B$ and D is the domain of x.

Note that if D is a finite set, then (5) and (6) become

(5)' $T(S) = T(B(x_1) \cdot \ldots \cdot B(a_n))$ if $S = (x)B$ and x assumes values $a_1, \ldots, a_n$.

(6)' $T(S) = T(B(a_1) + \ldots + B(a_n))$ if $S = (Ex)B$ and x assumed values $a_1, \ldots, a_n$.

The reader should note that two-valued logic is a special case of fuzzy logic; all the rules stated above are applicable in two-valued logic.

Example 2.7.1. Consider $S = (P + Q) \cdot (\overline{R})$. Assume $T(P) = 0.1$, $T(Q) = 0.7$, and $T(R) = 0.6$. Then

$$T(S) = \text{Min}\{\text{Max}[T(P), T(Q)], 1 - T(R)\}$$

$$= \text{Min}[\text{Max}(0.1, 0.7), 1 - 0.6]$$

$$= \text{Min}(0.7, 0.4)$$

$$= 0.4.$$

Example 2.7.2. Consider $S = (x)P(x)$. Assume the domain of x is (a_1, a_2, a_3) and $T(P(a_1)) = 0.1$, $T(P(a_2)) = 0.7$, and $T(P(a_3)) = 0.5$. Then

$$T(S) = T(P(a_1) \cdot P(a_2) \cdot P(a_3))$$

$$= \text{Min}[P(a_1), P(a_2), P(a_3)]$$

$$= \text{Min}(0.1, 0.7, 0.5)$$

$$= 0.1$$

Example 2.7.3. Consider $S = (x)(Ey)P(x,y)$. Assume $D = \{a_1, a_2\}$ and $T(P(a_1, a_1)) = 0.1$, $T(P(a_1, a_2)) = 0.3$,

$$T(P(a_2, a_1)) = 0.2, \qquad T(P(a_2, a_2)) = 0.7.$$

Note that

$$T((Ey)P(x,y)) = T(P(x,a_1) + P(x,a_2)).$$

Therefore

$$T\big((x)(Ey)P(x,y)\big) = T\big((P(a_1,a_1) + P(a_1,a_2)) \cdot (P(a_2,a_1) + P(a_2,a_2))\big)$$
$$= \text{Min}\{\text{Max}\big[P(a_1,a_1),P(a_1,a_2)\big],$$
$$\text{Max}\big[P(a_2,a_1),P(a_2,a_2)\big]\}$$
$$= \text{Min}\big[\text{Max}(0.1,\ 0.3),\ \text{Max}(0.2,\ 0.7)\big]$$
$$= \text{Min}(0.3,\ 0.7)$$
$$= 0.3.$$

If two-valued logic is used in a problem-solving system, one stores a statement A, instead of $\bar{A}$, if the truth value of A is 1. (If the truth value of a statement A is 0, one simply stores $\bar{A}$.) In fuzzy logic we store a statement A, instead of $\bar{A}$, if the truth value of A is greater than or equal to that of $\bar{A}$. That is, we store A if

$$T(A) \geqslant 1 - T(A).$$

In this case $T(A) \geqslant 0.5$.

We can use this concept to define "satisfiability" in fuzzy logic. An interpretation I is said to satisfy a formula S if $T(S) \geqslant 0.5$ under I. An interpretation I is said to falsify S if $T(S) \leqslant 0.5$ under I. (According to this definition, if $T(S) = 0.5$ under I, then I both satisfies and falsifies S.) A formula is said to be unsatisfiable if it is falsified by every interpretation of it. Again it should be easy to note that these definitions are compatible with those in two-valued logic. However, in fuzzy logic "not satisfying" is different from "falsifying," and "not falsifying" is different from "satisfying."

In two-valued logic a very important concept is the consistency of a formula. A formula A is said to be valid iff $T(A) = 1$ under all of its possible two-valued assignments. Similarly, a formula A is said to be inconsistent iff $T(A) = 0$ under all of its possible two-valued assignments. A formula is said to be consistent iff it is not inconsistent. A formula is said to be invalid iff it is not valid. In fuzzy logic we shall define similar concepts. A fuzzy formula is said to be fuzzily valid iff $T(A) \geqslant 0.5$ under all possible assignments and A is fuzzily inconsistent iff $T(A) \leqslant 0.5$ under all possible assignments.

Having defined validity and inconsistency of formulas in fuzzy logic, we can now investigate ways to determine consistency of formulas. In two-valued logic, since there are a finite number of distinct assignments for a formula, the most straightforward way to determine the consistency of a formula is an exhaustive search. That is, given a formula S, we examine every possible two-valued assignment. If S is 1(0) under all these assignments, we say that S is valid (inconsistent). However, we cannot have an exhaustive examination of all possible assignments in fuzzy logic because there are an infinite number of possible assignments for every formula. Therefore, some non-exhaustive way to determine consistency of formulas in fuzzy logic is very important.

To check whether a formula A is valid or inconsistent in fuzzy logic the simplest approach involves expanding A into conjunctive and disjunctive forms. Before defining these normal forms we give some definitions.

A literal is a variable X_i or $\overline{X_i}$, the complement of X_i.

A clause is a disjunction of one or more than one literal.

A phrase is a conjunction of one or more than one literal.

A formula A is said to be in conjunctive normal form if

$A = C_1 \cdot C_2 \cdot \ldots \cdot C_m$, $m \geq 1$ and every C_i, $1 \leq i \leq m$, is a clause.

A formula A is said to be in disjunctive normal form if

$A = P_1 + P_2 + \ldots + P_m$, $m \geq 1$ and every P_i, $1 \leq i \leq m$, is a phrase.

In two-valued logic it can be shown that every formula can be expressed in conjunctive and disjunctive normal forms owing to the existence of the distributive laws and De Morgan's laws. Since both of the laws mentioned above hold in fuzzy logic and since there is no syntactical difference between formulas in fuzzy logic and formulas in two-valued logic, we can easily see that formulas in fuzzy logic can also be expressed in conjunctive and disjunctive normal form.

Lemma 2.7.1. Let C be a clause. If C contains a complementary pair of literals then C is fuzzily valid.

Proof. Let $C = L_1 + L_2 \ldots + L_n$. Assume L_i and L_j form such a complementary pair. Then $T(L_i) = 1 - T(L_j)$. For every possible assignment, either $T(L_i)$ or $T(L_j)$ will be greater than or equal to 0.5. Therefore, $\text{Max}[T(L_i), T(L_j)] \geq 0.5$ for all possible assignments. Since

$$T(C) = \text{Max}[T(L_1), T(L_2), \ldots, T(L_n)]$$

$$T(C) = \text{Max}\big[T(L_1), T(L_2), \ldots, T(L_i), \ldots, T(L_j), \ldots, T(L_n)\big]$$
$$\geq \text{Max}\big[T(L_i), T(L_j)\big] \geq 0.5.$$

Thus C is fuzzily valid. Q.E.D.

Lemma 2.7.2. Let C be a clause. If C is fuzzily valid, then C contains a complementary pair of literals.

Proof. Consider the assignment in which every literal of C is assigned a truth value smaller than 0.5. $T(C)$ will be smaller than 0.5 under this assignment, so C is not fuzzily valid. This is contradictory to the assumption that C is fuzzily valid Q.E.D.

Combining Lemmas 2.7.1 and 2.7.2, we have the following theorem.

Theorem 2.7.1. Let C be a clause. C is fuzzily valid iff C contains a complementary pair of literals.

Similarly, we can prove the following theorem concerning inconsistency in fuzzy logic.

Theorem 2.7.2. Let P be a phrase. P is fuzzily inconsistent iff P contains a complementary pair of literals.

Theorems 2.7.1 and 2.7.2 can be utilized to check the consistency of formulas in fuzzy logic. Suppose we want to see whether a formula A is fuzzily valid. We can expand A into a conjunctive normal form:

$$A = C_1 \cdot C_2 \cdot \ldots \cdot C_n.$$

Then A is fuzzily valid iff every C_i is fuzzily valid. However, the fuzzy validity of a clause can be established through Theorem 1. Similarly, in case we want to check the fuzzy inconsistency of formula A, we can expand A into disjunctive normal form:

$$A = P_1 + P_2 + \ldots + P_n.$$

Then A is fuzzily inconsistent iff every phrase P_i is fuzzily inconsistent and the fuzzy inconsistency of a phrase can be established through Theorem 2.7.2.

2.8. FUZZY ALGEBRA

Intuitively, a fuzzy set is a class which admits the possibility of partial membership in it. Let $X = \{x\}$ denote a space of objects. Then a fuzzy set A in X is a set of ordered pairs

$$A = \{(x, \mu_A(x))\}, x \in X$$

where $\mu_A(x)$ is termed the grade of membership of x in A. We shall assume for simplicity that $\mu_A(x)$ is a number in the interval $[0,1]$ with the grades 1 and 0 representing respectively full membership and non-membership in a fuzzy set.

Definition 2.8.1. A *fuzzy algebra* is a system

$$Z = \langle Z, +, *, ^- \rangle$$

where Z has at least two distinct elements and $\forall\, x,y,z \in Z$; system Z satisfies the following set of axioms.

(1) Idempotency: $x + x = x$ $x * x = x.$

(2) Commutativity: $x + y = y + x$ $x * y = y * x.$

(3) Associativity: $(x + y) + z = x + (y + z)$ $(x * y) * z = x * (y * z).$

(4) Absorption: $x + (x * y) = x$ $x * (x + y) = x.$

(5) Distributivity: $x + (y * z) = (x + y) * (x + z)$ $x * (y + z) = x * y + x * z.$

(6) Complement: If $x \in Z$ then there is a unique complement $\bar{x}$ of x such that $\bar{x} \in Z$ and $\bar{\bar{x}} = x.$

(7) Identities: $(\exists! e_+)(\forall\, x)$ such that $x + e_+ = e_+ + x = x$

                    $(\exists! e_*)(\forall\, x)$ such that $x * e_* = e_* * x = x.$

(8) De Morgan Laws: $\overline{x + y} = \bar{x} * \bar{y}$ $\overline{x * y} = \bar{x} + \bar{y}.$

Clearly the system is a distributive lattice with the existence of unique identities under $+$ and $*$. It is noted that a Boolean algebra is a complemented distributive lattice with existence of unique identities under $+$ and $*$. However, for every element x in Boolean algebra there exists $\bar{x}$ such that $x\bar{x} = 0$ and $x + \bar{x} = 1$ which is not so in fuzzy algebra. Hence, every Boolean algebra is a fuzzy algebra, but not *vice versa*.

In this book we shall use a particular fuzzy algebra defined by the system

$$Z = \langle [0,1], +, *, ^- \rangle$$

where $+$, $*$, and $^-$ are interpreted as Max, Min, and complement ($\bar{x} = 1 - x$, $\forall\, x \in [0,1]$), respectively. The unique identities e_+ and e_* are 0 and 1, respectively.

In this fuzzy algebra the elements 0, 1 satisfy, for every fuzzy element x,

$$x + 0 = x \qquad x * 0 = 0$$
$$x + 1 = 1 \qquad x * 1 = x. \tag{2.8.1}$$

In the sequel, the term "fuzzy variable" will replace the term membership grade of a variable in a set. Conventionally, we shall drop the $*$ symbol, i.e., $x * y$ will be written xy.

We can now define fuzzy forms, generated by $x_1, \ldots, x_n$, recursively as follows.

(a) The numbers 0 and 1 are fuzzy forms.

(b) A fuzzy variable x_i is a fuzzy form.

(c) If A is a fuzzy form, then $\overline{A}$ is a fuzzy form.

(d) If A and B are fuzzy forms, then $A + B$ and AB are fuzzy forms.

(e) The only fuzzy forms are those given by rules (a)–(d).

The grade membership $\mu(S)$ of a form S is uniquely determined through the following rules.

(1) $\mu(S) = 0$ if $S = 0$.

(2) $\mu(S) = 1$ if $S = 1$.

(3) $\mu(S) = \mu(x_i)$ if $S = x_i$.

(4) $\mu(S) = 1 - \mu(A)$ if $S = \overline{A}$.

(5) $\mu(S) = \text{Min}[\mu(A), \mu(B)]$ if $S = AB$.

(6) $\mu(S) = \text{Max}[\mu(A), \mu(B)]$ if $S = A + B$.

Example 2.8.1. Let

$$f(x_1, x_2, x_3) = x_1(x_2 + x_3)(\bar{x}_2 + x_3)$$

and let the grade memberships of x_1, x_2, and x_3 be $\mu(x_1) = 0.4$, $\mu(x_2) = 0.7$, and $\mu(x_3) = 0.6$. Then

$$\begin{aligned}
\mu[f(x_1, x_2, x_3)] &= \text{Min}[\mu(x_1), \mu(x_2 + x_3), \mu(\bar{x}_2 + x_3)] \\
&= \text{Min}\{\mu(x_1), \text{Max}[\mu(x_2), \mu(x_3)], \text{Max}[\mu(\bar{x}_2), \mu(x_3)]\} \\
&= \text{Min}\{0.4, \text{Max}(0.7, 0.6), \text{Max}[1 - \mu(x_2), \mu(x_3)]\} \\
&= \text{Min}[0.4, 0.7, \text{Max}(1 - 0.7, 0.6)] \\
&= \text{Min}(0.4, 0.7, 0.6) = 0.4.
\end{aligned}$$

It is quite clear that among the infinite number of distinct assignments of grade membership to the variables, there are a finite number of binary assignments (binary assignments of 0 or 1 to every variable).

It can be shown that the set of fuzzy functions compatible with a given Boolean function $F_B(x)$ is a sublattice of the lattice of fuzzy functions. It can also be shown that the "soft" algebra is a bounded, distributive, and symmetric lattice. Hence it is clear that a finite fuzzy lattice is isomorphic to a ring of sets. In the case of Boolean lattices it is easy to prove that a finite lattice is Boolean if and only if it is isomorphic to the Boolean lattice of all subsets of a finite set. In any lattice we have the standard definition that an element ξ is an *atom* if $\xi \succ\!\!- 0$ (ξ covers 0) and a dual atom if $\xi -\!\!\prec 1$ (1 covers ξ), and thus all the elements which are immediate successors of the lower bound 0 are the atoms of the lattice.

Regarding Boolean algebra, it is obvious that the above definition of an atom and the classicial definition that a non-zero element ρ in a *Boolean algebra B* is a atom iff $\rho x = \rho$ or $\rho x = 0$ for every x in B are identical.

Because the n-variable switching functions form a Boolean algebra of order 2^{2^n}, the algebra has 2^n atoms which are just the minterms. In fuzzy algebra, however, we prefer the original definition of an atom which implies that the atom is the *minimal* non-zero element of the lattice. Thus an element ρ is an atom of the fuzzy lattice L iff $\rho \neq 0$ and for every x, if $x \leq \rho$, then $x = \rho$ or $x = 0$.

In order for an element to be the minimal element of L it must be the conjunction of all fuzzy variables and their complements, and thus ensure the minimal grade membership required. Hence it is clear that the fuzzy lattice has only a *single atom* which has the form of $\prod_j x_j \bar{x}_j$. The fuzzy algebra over n variables $(x_1, x_2, \ldots, x_n)$ has as its atom the element $x_1 \bar{x}_1 x_2 \bar{x}_2 \ldots x_j \bar{x}_j \ldots x_n \bar{x}_n$. This notion is a fuzzy analog of a one-point set. A fuzzy algebra Z is said to be *atomic* provided, for every non-zero element y in Z, there exists the atom $\prod_j x_j \bar{x}_j \leq y$. In a similar way it is clear that in a fuzzy algebra over n variables there exists a set of 2^n dual atoms which consists of the fuzzy variables and their complements.

2.9. ENTROPY OF A FUZZY SET

Let us consider a set I and a lattice L; any map from I to L is called an L-fuzzy set. Let us denote by $L(I)$ the class of all maps from I to L. It is possible to induce a lattice structure to $L(I)$ by the binary operations $\bigvee$ and $\bigwedge$ associating to any pair of elements f and g of $L(I)$ the elements $f \bigvee g$ and $f \bigwedge g$ of $L(I)$, defined point by point as

$$(f \vee g)(x) \equiv \text{l.u.b.}\{f(x), g(x)\}$$
$$(f \wedge g)(x) \equiv \text{g.l.b.}\{f(x), g(x)\} \tag{2.9.1}$$

where l.u.b. and g.l.b. denote respectively the least upper bound and the greatest lower bound of $f(x)$ and $g(x)$ in the lattice L.

In this paper we shall consider L as coinciding with the unit interval on the real line $L \equiv [0,1]$; in this case equation (2.9.1) becomes

$$(f \vee g)(x) = \text{Max}\{f(x), g(x)\}$$
$$(f \wedge g)(x) = \text{Min}\{f(x), g(x)\}. \tag{2.9.2}$$

We try to introduce, for every element, or "fuzzy set" $f \in L(I)$, a measure of the degree of its "fuzziness." We require of this quantity, which we shall denote by $d(f)$, that it must depend only on the values assumed by f on I and satisfy at least the following properties.

P1: $d(f)$ must be 0 if and only if f takes on I the values 0 or 1.

P2: $d(f)$ must assume the maximum value if and only if f assumes always the value $\frac{1}{2}$.

P3: $d(f)$ must be greater or equal to $d(f^*)$ where f^* is any "sharpened" version of f, that is any fuzzy set such that $f^*(x) \geq f(x)$ if $f(x) \geq \frac{1}{2}$ and $f^*(x) \leq f(x)$ if $f(x) \leq \frac{1}{2}$.

Let I be a finite set; this assumption and some others that we will make in the following simplify the mathematical formalism but may be suitably weakened in future generalizations, We note, however, that the finiteness of I corresponds to a large class of actual situations.

We begin by introducing on $L(I)$ the functional $H(f)$, formally similar to the Shannon entropy although quite different conceptually, whose range is the set of non-negative real numbers and defined as

$$H(f) \equiv -K \sum_{i=1}^{N} f(x_i) \ln f(x_i) \tag{2.9.3}$$

where N is the number of elements of I and K is a positive constant. We have the following:

Theorem 2.9.1. $H(f)$ is a non-negative valuation on the lattice $L(I)$, i.e. $H(f \vee g) + H(f \wedge g) = H(f) + H(g)$ for all f, g of $L(I)$. (2.9.4)

In fact, from definition (2.9.3) and by (2.9.2) it follows that

$$H(f \vee g) = -K \sum_{i=1}^{N} \max[f(x_i),g(x_i)] \ \ln \max[f(x_i),g(x_i)]$$

$$H(f \wedge g) = -K \sum_{i=1}^{N} \min[f(x_i),g(x_i)] \ \ln \min[f(x_i),g(x_i)].$$

(2.9.5)

By breaking up the sums in equation (2.9.5) into two parts, one extended over all x such that $f(x) \geqslant g(x)$ and the other over all x such that $f(x) < g(x)$, and summing up the right-hand and left-hand sides of equation (2.9.5), equation (2.9.4) is obtained.

It seems more convenient to us to introduce the following functional, which we shall call the "entropy" of the fuzzy set f,

$$d(f) \equiv H(f) + H(\bar{f}) \tag{2.9.6}$$

where $\bar{f}$, defined point by point as

$$\bar{f}(x) \equiv 1 - f(x)$$

satisfies the following noteworthy properties:

$$\bar{\bar{f}} = f \qquad \text{(involution law)}$$

$$\overline{f \vee g} = \bar{f} \wedge \bar{g}$$
$$\overline{f \wedge g} = \bar{f} \vee \bar{g}.$$
(De Morgan laws).

(2.9.7)

We explicitly note that $\bar{f}$, usually called the complement of f, is not the algebraic complement of f with respect to the lattice operations.

Clearly, $d(f) = d(\bar{f})$; moreover, $d(f)$ can be written using Shannon's function

$$S(x) = -x \ln x - (1 - x) \ln (1 - x) \text{ as}$$

$$d(f) = K \sum_{h=1}^{N} S(f(x_h)) \tag{2.9.8}$$

and $d(f)$ satisfies requirements P1 and P2. Requirement P3 is also satisfied. In fact, if f^* is a sharpened version of f we have by definition

(a) $0 \leqslant f^*(x) \leqslant f(x) \leqslant \frac{1}{2}, \quad$ for $0 \leqslant f(x) \leqslant \frac{1}{2}$

(b) $1 \geqslant f^*(x) \geqslant f(x) \geqslant \frac{1}{2}, \quad$ for $\frac{1}{2} \leqslant f(x) \leqslant 1$.

By the well-known property of Shannon's function $S(x)$, monotonically increasing in the interval $[0,\frac{1}{2}]$ and monotonically decreasing in $[\frac{1}{2},1]$ with a maximum at $x = \frac{1}{2}$, we immediately obtain from (a) and (b) that, for any value of $f(x)$,

$$S(f^*(x)) \leqslant S(f(x)), \quad x \in I.$$

From this relation by (2.9.8) it follows that

$$d(f^*) \leqslant d(f).$$

We can now prove the following result.

Theorem 2.9.2. $d(f)$ *is a non-negative valuation on the lattice* $L(I)$. *In fact we have*

$$d(f) + d(g) = H(f \vee g) + H(f \wedge g) + H(\bar{f} \wedge \bar{g}) + H(\bar{f} \vee \bar{g})$$
$$= H(f \vee g) + H(\overline{f \vee g}) + H(f \wedge g) + H(\overline{f \wedge g})$$
$$+ d(f \vee g) + d(f \wedge g).$$

If we assume in equation (2.9.8) that $K = 1/N$, we obtain the functional

$$v(f) = \frac{1}{N} \sum_{h=1}^{N} S(f(x_h)) \tag{2.9.9}$$

which we shall call the "normalized entropy." This name is appropriate because, taking the logarithm in base 2, one has

$$0 \leqslant v(f) \leqslant 1 \quad \text{for all } f \in L(I).$$

By Theorem (2.9.2) it follows immediatley that $v(f)$ is also a non-negative valuation on the lattice $L(I)$.

The functional $d(f)$ has been assumed as giving a measure of the fuzziness of f; this quantity, as we shall see, may also be considered as measuring an amount of information even if its meaning is different from the standard one of Shannon's information theory.

Let us now discuss the following example. We consider N cells x_i $(i = 1, \ldots, N)$ of sensory units (*e.g.*, photoelectric cells) disposed in a two-dimensional array I, or retina, and suppose first that one may project on the retina only patterns such that any cell can "see" only white or black colors, to which correspond two different states 1 or 0, of the photoelectric unit. Therefore, we may associate with the x_i cell $(i = 1, \ldots, N)$ a variable $f(x_i)$ which may assume only two values, 1 (white) and 0 (black). In this way, to any pattern corresponds a subset of I, that is, the one formed by the cells such that $f(x_i) = 1$.

Let us now suppose that for some reasons (generally depending on the projected pattern), whose nature at the moment does not interest us, the state $f(x_i)$ of the ith cell can vary in the interval $[0,1]$ instead of in the set $\{0,1\}$; meaning that the cell may "see" a discrete or continuous number of grey colors with each of which, by means of the photo-electric units, we associate a number that we interpret, according to the scale, as "degree of white" or "degree of black." In such a way a description of a pattern projected on the retina may be made by means of a fuzzy set f. This pattern, described by f, looks "ambiguous" to any "person" or "device" which knows only black or white; a meas-ure of this ambiguity is $d(f)$. The nature of this ambiguity therefore arises from the "incertitude" present when we must decide, looking at the grey color of the ith cell, if this has to be considered white or black. We may measure this incertitude by $S\left(f(x_i)\right)$, which is 0 if $f(x_i)$ is equal to 0 or 1 and is a maximum for $f(x_i) = \frac{1}{2}$; the total amount of incertitude is $\sum_{i=1}^{N} S\left(f(x_i)\right) = d(f)$.

Now if we carry out some experiments by which we can remove or reduce the uncertainty which existed before the experiment, we can say that we obtain some information. Let us, first, assume that the experiment consists in taking a decision about the colors (white or black) of all the N cells of the retina. In this way we produce a new classical pattern f^*. This kind of experiment makes the ambiguity of the final pattern f^* equal to 0 by completely removing the uncertainty on the colors of the cells which existed before the experiment. It seems natural to us to assume that in these experiments we receive an average amount of information proportional or equal to (choosing some unit) the initial uncertainty $d(f)$. We may assume that $d(f)$ also measures the average amount of information (about the colors of the pattern) which is lost going from a classical pattern to the fuzzy pattern f. It is also possible to consider only partial removals of uncertainty in any exper-iment by which we transform the fuzzy pattern $\bar{f}$ into a new fuzzy pattern $\bar{f}$ having

$$d(\bar{f}) \leq d(f).$$

In this way we may say that we receive a quantity of information measured by $d(f) - d(\bar{f})$. We emphasize that the ambiguity we have previously defined and the related information are "structural," that is, linked to the fuzzy description, while in classical information theory it is due to the uncertainty in the previsions of the results of random experiments.

Let us now consider any experiment in which the elements $x_1, \ldots, x_N$ of I may occur, one and only one in each trial, with prob-abilities $P_1, P_2, \ldots, P_N$ ($P_i \geq 0$, $\sum_{i=1}^{N} P_i = 1$). If a fuzzy set f is defined in I, we have two kinds of uncertainty.

(i) The first uncertainty of "random" nature is related to the prevision of the result, i.e. the element of I which will occur. As is well known, the average uncertainty is measurable by Shannon's entropy

$$H(P_1, \ldots, P_N) = -\sum_{i=1}^{N} P_i \ln P_i \; ;$$

H also gives the average information which is received knowing the element which occurs.

(ii) The second uncertainty of "fuzzy" nature concerns the interpretation of the result as 1 or 0. If the result is x_i we still have an amount of incertitude measurable by

$$S(f(x_i)).$$

The statistical average of $S(f(x_i))$,

$$m(f, P_i, \ldots, P_N) = \sum_{i=1}^{N} P_i S(f(x_i)) \tag{2.9.10}$$

which coincides with the normalized entropy (equation (2.9.9)) if $P_1 = P_2 = \ldots = P_N = 1/N$ represents the (statistical) average information received taking a decision (1 or 0) on the elements x_i ($i = 1 \ldots N$). This is an interesting new concept because it may happen, for instance, that the elements of I with $f(x) \approx \frac{1}{2}$ may occur in the random experiment only exceptionally where the elements with $f(x)$ near the bounds 1 or 0 may occur very frequently. In such a case m is small even if f is quite "soft"; this happens because the statistical uncertainty on the decisions is, in fact, small.

We can consider the total entropy as

$$H_{\text{tot}} = H(P_1 \ldots P_N) + m(f, P_1 \ldots P_N), \tag{2.9.11}$$

which may be interpreted as the total average uncertainty that we have in making a prevision about the element of I which will occur as a result of the random experiment and in taking a decision about the value 1 or 0 which has to be attached to the element itself. If $m = 0$, which occurs in the absence of fuzziness, H_{tot} reduces itself to the classical entropy $H(P_1 \ldots P_N)$. If we have $H(P_1 \ldots P_N) = 0$, which means there is no random experiment and only a fixed element, say x_i, will occur, then $H_{\text{tot}} = S(f(x_i))$.

We observe that the previous formula (2.9.11) is formally identical to one of the ordinary information theory

$$H(AB) = H(A) + H_A(B)$$

giving the entropy of the product scheme AB in the case in which the events of B are statistically dependent on those of A.

Another case which we may consider is when the fuzzy set f is random; that is f is a map

$$f: \Omega \times I \to [0,1]$$

such that, for any fixed x, $f(\xi,x)$ is a random variable with respect to a given probability space (Ω,F,p) where Ω is the non-empty set of sample points, F a σ-field of subsets of Ω, and p a probability measure. For any fixed ξ, $f(\xi,x)$ is a fuzzy set. Let us consider the case when Ω has only a finite number M of elements $\xi_1,\dots,\xi_M$, which may occur with probabilities $p(\xi_1),\dots,p(\xi_M)$; we may introduce an average fuzzy set $\langle f \rangle$ as

$$\langle f(x) \rangle \equiv \sum_{i=1}^{M} f(x,\xi_i)\, p(\xi_i).$$

In such a case the entropy of the fuzzy set is itself a random variable; in fact, if the event ξ_i happens, we have the fuzzy set $f(\xi_i,x)$ whose entropy is $d_i(f)$. In this case it is meaningful to consider the average entropy given by

$$\sum_{i=1}^{M} p(\xi_i)\, d_i = \sum_{i=1}^{M} \sum_{j=1}^{N} p(\xi_i)\, S\big(f(\xi_i,x_j)\big) \qquad (2.9.12)$$

which reduces itself to (2.9.8) in the deterministic case.

2.10. SOME ALGEBRAIC PROPERTIES OF FUZZY SETS

In this section we shall synthesize some relevant algebraic properties of fuzzy sets and point out new algebraic aspects of the theory, connecting it with brouwerian lattices. We recall some preliminary definitions and properties. Let I be a (non-empty) universal class whose general element is denoted by x. We assume the following definition of *fuzzy set* on I

(1) *A fuzzy set (on I) is a map*

$$f : I \to L$$

where L is a partially ordered set (poset).

We shall denote by $L(I)$ the class of all maps from I to L. If L consists of two elements only, $L \equiv \{0,1\}$, then each fuzzy set is an ordinary *characteristic function* defining an ordinary subset of I. If L is the closed interval $[0,1]$ of the real field we have the *generalized characteristic functions* or fuzzy sets.

The most interesting posets that may be considered are the *lattices*. In such a case the following proposition holds.

(2)*If L is a lattice, then so is $L(I)$ with respect to the operations* $\vee$ *(join) and* $\wedge$ *(meet) defined as follows.*

$$(f \vee g)(x) = \text{l.u.b. } \{f(x), g(x)\}, \text{ for all } x \in I$$
$$(f \wedge g)(x) = \text{g.l.b. } \{f(x), g(x)\}, \text{ for all } x \in I$$
(2.10.1)

In $L(I)$ the partial order relation $\leq$ is then defined as

$$f \leq g \Leftrightarrow f(x) \leq g(x), \text{ for all } x \in I$$
(2.10.2)

having

$$f = g \Leftrightarrow f(x) = g(x), \text{ for all } x \in I.$$

(3) *More generally any binary operation* $(\cdot)$ *defined on L can be induced point by point on $L(I)$ by:*

$$(f \cdot g)(x) = f(x) \cdot g(x), \text{ for all } x \in I.$$
(2.10.3)

The same is, of course, also true for unary or generally for n-ary operations.

Confining ourselves to the case of posets no more general than lattices, we emphasize that the most relevant algebraic properties of the lattice L are induced on $L(I)$ by the lattice operations (2.10.1). In particular one may easily see the following.

(4) *The modularity or the distributivity of L is transmitted on $L(I)$; moreover if L is a complete, complemented or Boolean lattice, then so is $L(I)$.*

We explicitly note that not every property of L can be used in $L(I)$ by (2.10.1) if L is a chain, for instance, $L(I)$ is not so (except the pathological case in which I consists of only one element).

Let us now assume L to be a *brouwerian lattice*. The following noteworthy property holds.

(5) *If L is a brouwerian lattice then so is $L(I)$.*

To prove this we have to show that for any pair of elements f and g of (I) the set of all functions of $L(I)$, such that

$$f \wedge \psi \leq g$$
(2.10.4)

contains a greatest element f^g called relative pseudo-complement of f in g. We can construct f^g defining, for any x, $(f^g)(x)$ as the greatest element (that by hypothesis exists) of the set of the elements z of L satisfying the relation

$$f(x) \wedge z \leqslant g(x).$$

If we introduce on $L(I)$ the binary operation $\div$

$$g \div f = \text{Def } f^g, \text{ for all } f,g \in L(I)$$

it is easily seen, by the definition of brouwerian lattice, that

$$f \geqslant g \wedge (f \div g)$$

$$(f \wedge g) \div g \geqslant f$$

$$f \div h \geqslant (f \wedge g) \div h.$$

Conversely if $L(I)$ is closed with respect to a binary operation such that the previous relations are satisfied then $L(I)$ is a brouwerian lattice having

$$f^g = g \div f.$$

If L is a chain with 0 and 1, then $L(I)$ will be a non-complemented lattice, whose only complemented elements are the classical characteristic functions forming a narrow sublattice of $L(I)$. However, *since a chain with 1 is a brouwerian lattice, such will be, by* (5), $L(I)$. In this case the *relative pseudo-complement of f on g* is the function defined, for any x, as

$$f^g(x) = \begin{cases} 1, & \text{if } f(x) \leqslant g(x) \\ g(x), & \text{if } f(x) > g(x). \end{cases}$$

If $g = 0$ the *pseudo-complement* (or *brouwerian complement*) of f, e.g. f^0, usually denoted by f^*, is given by

$$f^*(x) = \begin{cases} 1, & \text{if } f(x) \leqslant 0 \\ 0, & \text{if } f(x) > 0. \end{cases}$$

Similarly, one can prove that, *if L is a complete brouwerian lattice (as a chain with 0 and 1), $L(I)$ will be a complete brouwerian lattice having the complete distributivity of meet on joins, that is*

$$f \wedge (\bigvee g_\alpha) = \bigvee (f \wedge g_\alpha), \text{ for any set } \{g_\alpha\} \text{ and any } f. \qquad (2.10.5)$$

In the foregoing discussion we have attempted to convey some of the results behind the algebraic properties of fuzzy sets. For more details the reader is referred to related papers listed in the Bibliography.

Exercises

(1) "The union of fuzzy sets A and B is the smallest fuzzy set containing both A and B", which is also equivalent to "If D is any fuzzy set which contains both A and B, then it also contains the union of A and B." Prove the above statement.

(2) Similarly, prove that the definition of intersection of two fuzzy sets given in the text is equivalent to the following definition: "The intersection of fuzzy sets A and B is the largest fuzzy set which is contained in both A and B," which is also equivalent to "If D is any fuzzy set which is contained in both A and B, then it is also contained in the intersection of A and B."

(3) For any fuzzy set A prove that
  (a) $A \cup \emptyset = A$
  (b) $A \cap \emptyset = \emptyset$
  (c) $A \cup U = U$
  (d) $A \cap U = A$
  (e) $A \cup \overline{A} = U$
  (f) $A \cap \overline{A} = \emptyset$.

(4) The following relations hold for ordinary sets
  (a) $\emptyset - A = \emptyset$
  (b) $A - B = A \cap \overline{B}$
  (c) $A - (A - B) = A \cap B$
  (d) $C \cap D = A - [(A - C) \cup (A - D)]$
  (e) $C \cup D = A = [(A - C) \cap (A - D)]$.

If all the relative complement operations are changed to the symmetrical difference operation $\triangle$ defined in Definition 2.3.7, do the above relations hold for fuzzy sets?

(5) Let A and B be fuzzy sets. Prove that
$$A \triangle B = (A \cup B) \triangle (B \cap A).$$

(6) Prove that $A \odot B = \overline{(\overline{A} \, \overline{B})}$

(7) Show that $A°(B°C) = (A°B)°C$.

(8) Which of the following functions may describe a convex fuzzy set?

(a) $\sin x \cdot [\mu(x) - \mu(x - 1\pi) + (x - 2\pi) - (x - 3\pi)]$

(b) $|\cos x| \cdot [\frac{1}{4}\mu(x) + \frac{1}{2}\mu(x - \pi/2) + \frac{1}{4}\mu(x - 3\pi/2 - \mu(x - \pi/2)]$

(c) $(x/\cosh x) \cdot [\mu(x) - \mu(x - 1) - \mu(x - 2) + \mu(x - 3)]$.

(9) The following are some of the well-known continuous probability density functions.

The uniform distribution

$$f_U(x) = \frac{1}{a} [\mu(x - \alpha) - \mu(x - \alpha - a)]$$

The normal distribution

$$f_N(x) = \frac{1}{(2\pi\sigma)^{1/2}} \exp\left[\frac{-(x - \mu)^2}{2\sigma^2}\right], -\infty < x < \infty$$

The gamma distribution

$$f_\gamma(x) = \frac{1}{\alpha!\,\beta^{\alpha+1}} x^\alpha \exp\left(-\frac{x}{\beta}\right), \quad 0 < x < \infty$$

$$= 0 \qquad \text{elsewhere}$$

where $\alpha > -1$ and $\beta > 0$.

The beta distribution

$$f_\beta(x) = \frac{(\alpha + \beta + 1)!}{\alpha!\,\beta!} x^\alpha(1 - x)^\beta, 0 < x < 1$$

$$= 0 \qquad \text{elsewhere}$$

where α and β must both be greater than -1.

The Cauchy density function

$$f_C(x) = \frac{1}{\pi} \frac{a}{1 + a^2 (x - \mu)^2}, \qquad -\infty < x < \infty$$

Show that any fuzzy set described by one of these functions is convex.

(10) Let A be a fuzzy set with membership function $f_A(x)$. Define the sets

$$S_\alpha = \{x \mid f_A(x) \geq \alpha\}.$$

Prove that the definition of a convex set given in Definition 2.5.1. is equivalent to the following definition: "A fuzzy set A is convex if and only if the sets S_α are convex for all $\alpha \in [0,1]$"

(11) Let A,B be two convex fuzzy sets. Show, by a counter example, that the following fuzzy sets obtained by set operations are not necessarily convex.

(a) $A - B$

(b) $A \cup B$

(c) $A \triangle B$

(d) $A + B$

(e) $A \oplus B$

(12) In exercise 4, if we impose one additional condition $B \subseteq A$ on A and B, then which of the above operations preserve the convexity property?

(13) In exercise 4, if A and B are disjoint sets, which of the operations preserve the convexity property?

(14) A fuzzy set is *concave* if A is the complement of a convex set. More directly, we call a fuzzy set A in X *concave* if it satisfies the following condition:

$$F_A(\lambda x_1 + (1 - \lambda)x_2) \leq \text{Max} \left[f_A(x_1), f_A(x_2) \right]$$

for all $x_1, x_2 \in X$ and $\lambda \in [0,1]$. Prove that the concavity property is preserved under the operation of union.

(15) Let A be a convex fuzzy set in R'' and let $S_H^*(A)$ be an orthogonal shadow of A on a hyperplane H. Then $S_H^*(A)$ is a convex fuzzy set in H. Prove or disprove.

(16) Let A be a fuzzy set with membership function $f_A(x)$ and let k be a constant, $0 \leq k \leq 1$. Define kA to be a fuzzy set whose membership function is given by

$$f_{kA}(x) = kf_A(x).$$

Show that $S_H(kA) = kS_H(A)$ (homogeneity property).

(17) For any two fuzzy sets A and B, if $A \subseteq B$, show that

$$S_H(A) \subseteq S_H(B).$$

(18) For any two fuzzy sets A and B, prove that

$$S_H(A \cup B) = S_H(A) \cup S_H(B) \quad \text{(distributivity property)}.$$

More generally, for any constants k_1 and k_2 in $[0,1]$, show that

$$S_H(k_1 A \cup k_2 B) = k_1 S_H(A) \cup k_2 S_H(B).$$

(19) Show that in general $S_H(A \cap B) \neq S_H(A) \cap S_H(B)$. As a matter of fact, for any two fuzzy sets, we have

$$S_H(A \cap B) \subseteq S_H(A) \cap S_H(B)$$

(20) The *complementary shadow* $C_{H_i}(A)$ of a fuzzy set A on H_i is a fuzzy set whose membership function is given by

$$f_{C_{H_i}(A)}(x) = \operatorname*{Inf}_{p_i} f_A(p_1, p_2, \ldots, p_n), \quad x \in H_i$$

$$= 0 \qquad\qquad , \quad x \notin H_i.$$

Show that if A is concave, then so is $C_{H_i}(A)$.

(21) The *cylindrical* fuzzy sets $S^c_{\bar{H}_i}(A)$ and $C^c_{\bar{H}_i}(A)$ are fuzzy sets whose membership functions are given by

$$f_{S^c_{\bar{H}_i}(A)}(x) = \operatorname*{Sup}_{p_i} f_A(p_1, p_2, \ldots, p_n)$$

$$f_{C^c_{\bar{H}_i}(A)}(x) = \operatorname*{Inf}_{p_i} f_A(p_1, p_2, \ldots, p_n).$$

Show that A is bounded from above by the intersection of the $S^c_{\bar{H}_i}(A)$ and is bounded below to the union of the $C^c_{\bar{H}_i}(A)$. Symbolically,

$$\bigcup_{i=1}^{n} C^c_{\bar{H}_i}(A) \subseteq A \subseteq \bigcap_{i=1}^{n} S^c_{\bar{H}_i}(A).$$

(22) A die is cast repeatedly until each of the six faces appears at least once. Define a fuzzy event A as "the die is cast approximately 10 times," whose membership function is defined by

$$f_A(x) = \begin{cases} 0 & x < 6 \\ \dfrac{1}{1 + (x - 10)^2} & x \geq 6 \end{cases}$$

where x denotes the number of times the die is cast. What is the probability that it is cast approximately 10 times?

(23) A person repeatedly casts a pair of dice. Let A and B be fuzzy events defined by

A = a number which is close to 8 is cast

B = a number which is close to 5 is cast

which are respectively described:

$$f_A(x) = \exp\left(-|x - 8|\right)$$
$$f_B(x) = \exp\left(-2|x - 5|\right)$$

where x denotes the number he casts. He wins if he casts a number close to 8 before he casts a number close to 5. What is his approximate probability of winning?

(24) A city C has the following weather transition probability matrix:

		R	S	C	SN	ST	F
	R	$\frac{1}{8}$	$\frac{5}{16}$	$\frac{5}{16}$	$\frac{1}{16}$	$\frac{1}{16}$	$\frac{1}{8}$
	S	$\frac{1}{8}$	$\frac{7}{16}$	$\frac{3}{16}$	$\frac{1}{16}$	$\frac{1}{16}$	$\frac{1}{8}$
$P =$	C	$\frac{1}{2}$	$\frac{1}{16}$	$\frac{1}{16}$	$\frac{1}{4}$	$\frac{1}{16}$	$\frac{1}{16}$
	SN	$\frac{1}{16}$	$\frac{1}{16}$	$\frac{1}{16}$	$\frac{1}{2}$	$\frac{1}{16}$	$\frac{1}{4}$
	ST	$\frac{1}{2}$	$\frac{1}{16}$	$\frac{1}{16}$	$\frac{1}{4}$	$\frac{1}{16}$	$\frac{1}{16}$
	F	$\frac{1}{16}$	$\frac{7}{16}$	$\frac{3}{16}$	$\frac{1}{16}$	$\frac{1}{16}$	$\frac{3}{16}$

where R, S, C, SN, ST and F denote a rainy, sunny, cloudy, snowy, stormy, and foggy day, respectively. Thus the sample space $S = \{R, S, C, SN, ST, F\}$. Let A and B be two fuzzy events

A = a warm day
B = a wet day

which are subjectively defined by

f_A: $f_A(R) = 0.4$, $f_A(S) = 0.9$, $f_A(C) = 0.6$,
$\qquad f_A(SN) = 0.05$, $f_A(ST) = 0.3$, $f_A(F) = 0.5$

f_B: $f_B(R) = 0.9$, $f_B(S) = 0.1$, $f_B(C) = 0.2$,
$\qquad f_B(SN) = 0.75$, $f_B(ST) = 0.7$, $f_B(F) = 0.3$.

Suppose we start with a cloudy day.

   (a) What is the probability of having a warm day two days later?

   (b) What is the probability of having a wet day three days later?

   (c) What is the probability of having a cold day the following day?

   (d) What is the probability of having a warm or dry day two days later?

   (e) What is the probability of having a cold and wet day the following day?

(25) Suppose in a city a football team of m players is going to be formed. The coach is interested in selecting players according to their heights and weights. (Of course these characteristics are not enough in reality, but we consider just two characteristics here for simplicity.) Suppose he considers an ideal football player should be about 6 ft. tall and weigh about 200 lbs. He asks those who are interested in playing football and whose weight is close to 200 lbs. and whose height is close to 6 ft. to enrol. It turns out that there are $n(>m)$ men enrolled. The problem is: How should he select m men from the group of n men in such a way that the m men selected are closer to his standard than those remaining? Could you suggest a solution for him?

CHAPTER 3

Fuzzy Functions and Decomposition

3.1. INTRODUCTION

This chapter is concerned with the study of fuzzy switching functions in combinational switching systems by means of a suitable fuzzy algebra and techniques for their minimization and decomposition. A fuzzy function can usually be represented by several expressions. Our aim in this chapter will be to develop techniques for obtaining a minimal expression for any such function, after establishing some criteria for minimality. It is possible, of course, to simplify fuzzy expressions by means of algebraic manipulations. One takes the fuzzy expression and applies the identities of fuzzy algebra to obtain simplified forms. The deficiency of this approach is that it really does not constitute an algorithm and is ineffective for expressions of even a small number of variables. The methods to be presented in this chapter will be more systematic, and will expose the reader to various relationships between fuzzy functions and binary logic.

3.2. FUZZY FUNCTIONS AND NORMAL FORMS

Any fuzzy logic function can be realized, using the basic fuzzy logic AND, OR, and inverter circuits, by associating the multiplication factors to the variables at different stages.

The electronic implementation of fuzzy functions is easily carried out by utilizing, for instance, diode-transistor logic. The operations

$x \cdot y$ and $x + y$ have exactly the same electronic implementation on the AND and OR logic operations of two-valued logic.

The logic operation complementation is easily implemented by making use of a transistor inverter as shown in Fig. 3.1. In designing the inverter care must be exercised to ensure a set of input–output characteristics similar to those in Fig. 3.1b. These types of characteristics can be approximated fairly close by biasing the transistor properly. It is now apparent that when $V_x = 0$, which corresponds to a membership grade of $g = 0$, the output voltage $V_o = V_{cc} - V_x$, which is equivalent to $\bar{x} = 1 - x$.

Even though the electronic implementation of fuzzy functions is easily realized, the practical usefulness of the fuzzy function concept and its electronic implementation are very difficult issues to defend. In other words, a fuzzy function as such does not provide us with a "decision mechanism" in the sense that a two-valued (or n-valued) logic function does with its "one–zero", "yes–no", or true–false" states. To know, for instance, that a fuzzy function, $f(x,y)$ has the value $f(x,y) = 0.8$ does not have much significance unless it is related to some outcome which is dependent upon the fact that $f(x,y) = 0.8$. A possible way for introducing this quality into a fuzzy system is by reintroducing the abandoned concept of "belonging to a set" but in a rather qualified sense. That is, we agree to have the continuous range of membership grades $[0,1]$ subdivided into a finite number of classes in the manner shown below.

Class 1: $\alpha_1 \leqslant x \leqslant 1$

Class 2: $\alpha_2 < x < \alpha_1$ (3.2.1)
$\vdots$ $\vdots$

Class n: $0 \leqslant x \leqslant \alpha_{n-1}$

where $1 > \alpha_1 > \alpha_2 > \ldots > \alpha_{n-1} > 0$. A fuzzy variable or a fuzzy function can now be identified with one of these m classes according to the value they assume in the closed interval $[0,1]$. This subdivision of the region $[0,1]$ into a finite number of classes enables one to utilize the properties of m-valued logics in the treatment of fuzzy logic systems. For example, for $m = 3$,

Class 1: $\alpha_1 \leqslant x \leqslant 1$

Class 2: $\alpha_2 < x < \alpha_1$ (3.2.2)

Class 3: $0 \leqslant x \leqslant \alpha_2.$

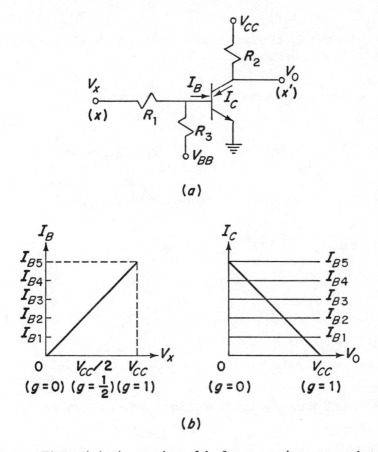

FIG. 3.1. Electronic implementations of the fuzzy operations $\cdot$, $+$, and $-$.

We may now ascribe certain meanings to the different classes. For instance, we may assume that an object x (a) belongs to a set if it is in "class 1", (b) does not belong to a set if it is in "class 3", and (c) its status remains undecided if it is in "class 2".

3.2.1. Analysis of Fuzzy Functions

In the analysis of fuzzy functions, we consider two functional representations: the "sum-of-products" and "product-of-sums" forms. Any other functional form would be a combination of these two forms.

(A) Sum-of-product form

The analysis of fuzzy functions of sum-of-product form is best illustrated by use of examples.

Example 3.2.1. Consider a simple fuzzy function

$$f(x,y) = x + \bar{y}. \tag{3.2.3}$$

In analyzing the function, one is interested in determining the conditions that must be satisfied by fuzzy variables x and y in order for the function to belong to a certain class M as defined in equation (3.2.1) above. Let $f(x,y)$ belong to class M or, equivalently, let

$$a_M \leq f(x \cdot y) < a_{M-1} \tag{3.2.4}$$

which is

$$x + \bar{y} \geq a_M \tag{3.2.4a}$$

and

$$x + \bar{y} < a_{M-1}. \tag{3.2.4b}$$

From equation (3.2.4a), we have

$$x > a_M \quad \text{or} \quad \bar{y} \geq a_M$$

which is equivalent to $y \leq 1 - a_M$. Taking the complement of both sides of equation (3.2.4b) and applying DeMorgan's law we obtain

$$\bar{x} \cdot y > 1 - a_{M-1} .$$

which means that $y > 1 - a_{M-1}$ and $\bar{x} > 1 - a_{M-1}$ which is equivalent to $x < a_{M-1}$.

In summary, in order for the function of equation (3.2.3) to belong to the class $a_M \leq f(x,y) < a_{M-1}$, x and y must satisfy the following conditions:

$$\text{Group } 1 = \left[x \geq a_M \quad \text{or} \quad y \leq 1 - a_M \right] \tag{3.2.5a}$$

and

$$\text{Group } 2 = \left[x < a_{M-1} \quad \text{and} \quad y > 1 - a_{M-1} \right] \tag{3.2.5b}$$

where group 1 is the set of conditions on x and y so that $f(x,y) \geq a_M$ and group 2 is the set of conditions on x and y so that $f(x,y) < a_{M-1}$.

Since taking the complement of both sides of the relation $f(x,y) < a_M$ results in $\bar{f}(x,y) > (1 - a_M)$, the set of the conditions obtained from this relation is dual with that obtained from $f(x,y) \geq a_M$. In other words, group 1 and group 2 are dual in nature.

Example 3.2.2. Suppose the fuzzy function is

$$f(x,y,z) = x\bar{y}z + y\bar{z}. \tag{3.2.6}$$

We want to find the conditions that must be satisfied by the fuzzy variables x, y, and z so that $a_M \leq f(x,y,z) < a_{M-1}$. From the condition $a_M \leq f(x,y,z)$ we immediately have

$$\text{Group 1} = \left[\left\{ \begin{array}{l} x \geq a_M \\ \text{and } y \leq 1 - a_M \\ \text{and } z \geq a_M \end{array} \right\} \text{ or } \left\{ \begin{array}{l} y \geq a_M \\ \text{and } z \leq 1 - a_M \end{array} \right\} \right] \tag{3.2.7}$$

and from the condition $f(x,y,z) < (1 - a_{M-1})$ we obtain

$$\text{Group 2} = \left[\left\{ \begin{array}{l} x < a_{M-1} \\ \text{or } y > 1 - a_{M-1} \\ \text{or } z < a_{M-1} \end{array} \right\} \text{ and } \left\{ \begin{array}{l} y < a_{M-1} \\ \text{or } z > 1 - a_{M-1} \end{array} \right\} \right] \tag{3.2.8}$$

(B) Product-of-sum form

Next consider the product-of-sum form.

Example 3.2.3. Consider a simple function in product-of-sum form:

$$f(x,y,z) = (x + \bar{y} + z)(\bar{x} + y + \bar{z}). \tag{3.2.9}$$

Following the same procedure presented in *(A)*, we have that in order for $f(x,y,z)$ to belong to class *M* or equivalently

$$a_M \leq f(x,y,z) < a_{M-1}$$

the fuzzy variables x, y, and z must satisfy the conditions given as

$$\text{Group 1} = \left[\left\{\begin{array}{l} x \geqslant a_M \\ \text{or } y \leqslant 1 - a_M \\ \text{or } z \geqslant a_M \end{array}\right\} \text{ and } \left\{\begin{array}{l} x \leqslant 1 - a_M \\ \text{or } y \geqslant a_M \\ \text{or } z \leqslant 1 - a_M \end{array}\right\}\right] \quad (3.2.10)$$

and

$$\text{Group 2} = \left[\left\{\begin{array}{l} x < a_{M-1} \\ \text{and } y > 1 - a_{M-1} \\ \text{and } z < a_{M-1} \end{array}\right\} \text{ and } \left\{\begin{array}{l} x \leqslant 1 - a_{M-1} \\ \text{or } y \geqslant a_{M-1} \\ \text{or } z \leqslant 1 - a_{M-1} \end{array}\right\}\right] \quad (3.2.11)$$

where groups 1 and 2 are defined as before.

From the above examples, we find that not only are groups 1 and 2 dual in nature, but the conditions on the variables for the two forms are also dual in nature. More specifically, the rules of constructing sets of conditions (groups 1 and 2) *directly* from the function which is either of the sum-of-product form or the product-of-sum form may be stated as follows. These rules hold for both forms.

(1) The fuzzy variables of group 1 associated with inequalities directed to the left appear in their complemented form in the term they enter, while those associated with inequalities directed to the right appear in their uncomplemented form.

(2) The fuzzy variables of group 2 associated with inequalities directed to the left appear in their uncomplemented form in the term they enter, while those associated with inequalities directed to the right appear in their complemented form.

(3) The connectives AND and OR represent the logic operations of intersection and union, respectively, when associated with group 1; the same connectives are logically reversed if associated with group 2. Thus, in the latter case, OR and AND mean intersection and union respectively.

Example 3.2.4. Finally, consider the fuzzy function

$$f(x,y,z,w) = xy(\bar{z} + \bar{w}) + \bar{x}\bar{y} + zw \quad (3.2.12)$$

which obviously does not belong to either one of the two forms examined previously. However, the sets of conditions groups 1 and 2 can

still be obtained by directly applying the rules stated above. The result
is

$$
\text{Group 1} = \left[\left\{\begin{array}{l} x \geqslant \alpha \\ \text{and } y \geqslant \alpha \\ \text{and } (z \leqslant 1 - \alpha \text{ or } w \leqslant 1 - \alpha) \end{array}\right\}\right]
$$

$$
\text{or} \quad \left[\left\{\begin{array}{l} x \leqslant 1 - \alpha \\ \text{and } y \leqslant 1 + \alpha \end{array}\right\} \quad \text{or} \quad \left\{\begin{array}{l} z \geqslant \alpha \\ \text{and } w \geqslant \alpha \end{array}\right\}\right]
$$

$$(3.2.13)$$

and

$$
\text{Group 2} = \left[\left\{\begin{array}{l} x < \alpha \\ \text{or } y < \alpha \\ \text{or } (z > 1 - \alpha \text{ and } w > 1 - \alpha) \end{array}\right\}\right]
$$

$$
\text{or} \quad \left[\left\{\begin{array}{l} x > 1 - \alpha \\ \text{and } y > 1 - \alpha \end{array}\right\} \quad \text{or} \quad \left\{\begin{array}{l} z < \alpha \\ \text{and } w < \alpha \end{array}\right\}\right]
$$

$$(3.2.14)$$

In a disjunctive normal from, each phrase corresponds to a logic
gate and each literal to an input line. The ratio between the cost of a
logic gate and the cost of an input line will depend on the type of gates
used in the realization. However, in practice the cost of an additional
input line on an already existing gate will be several times less than the
cost of an additional logic gate. On this basis the elimination of gates
will be the primary objective of the minimization process, leading to
the following definition of a mimimal expression.

Definition 3.2.1. A disjunctive normal form is regarded as a *minimal complexity form* if there exists (1) no other equivalent form involving fewer phrases, and (2) no other equivalent form involving the same
number of phrases but a smaller total number of literals.

Definition 3.2.2a. A phrase f_j subsumes another phrase f_i iff f_j contains all the literals of f_i. $[\mu(f_j) \leqslant \mu(f_i)]$.

Implicant and prime implicant of fuzzy functions may be defined by using the concept of containment in fuzzy sets.

Definition 3.2.2b. A fuzzy function F_1 *covers* another function F_2, denoted $F_1 \supseteq F_2$, iff the value that F_1 assumes is always greater than or equal to the value that F_2 assumes for all values of the domain. Let G be a product of literals; G is said to be an *implicant* of F_1 iff F_1 covers G. A *prime implicant* P of a fuzzy function F_1 is a product term (a product of literals) which is covered by F_1, such that the deletion of any literal from P results in a new product which is not covered by F_1.

Clearly, implicant and prime implicant are defined in Definition 3.2.2b without any restrictions on the form of representation (e.g. sum-of-products form or product-of-sums form) of a fuzzy function. Also, a phrase subsuming another phrase becomes a special case of definition 3.2.2b.

A fuzzy implicant f_j is said to be a *fuzzy prime implicant* (FPI) if it subsumes no other fuzzy implicant of F

(i.e. $\mu(f_j) \le \mu(f_k) \le \mu(F) \Leftrightarrow k = j$).

In Boolean algebra we apply the following identities in order to minimize a Boolean function:

$x + x = x$	$x \cdot x = x$	(3.2.15a)
$x + 0 = x$	$x \cdot 1 = x$	(3.2.15b)
$x + 1 = 1$	$x \cdot 0 = 0$	(3.2.15c)
$x + \bar{x} = 1$	$x \cdot \bar{x} = 0.$	(3.2.15d)

These identities can be applied to Boolean variables or Boolean functions, and together with distributivity and absorption they form the set of rules by which Boolean minimization is performed. It is clear that rules 1, 2, and 3 are also true in fuzzy algebra, and that fuzzy algebra satisfies distributivity and absorption.

In general, one cannot apply the identities in (3.2.15d) to fuzzy expressions, as can be easily seen from the following example.

Example 3.2.5. Let $F = x_1 + \bar{x}_1 x_2$. Clearly, in Boolean algebra

$$F = (x_1 + \bar{x}_1)(x_1 + x_2) = x_1 + x_2$$

by applying (3.2.15d). However, this is not so in fuzzy algebra (e.g. let $x_1 = 0.4$, $x_2 = 0.7$. This implies that $x_1 + x_2 = 0.7$ but

$$x_1 + \bar{x}_1 x_2 = \text{Max} \left[0.4, \text{Min} (0.6, 0.7) \right] = 0.6.$$

In certain cases the fuzzy version of identities (3.2.15d) can be applied. These cases are discussed in the following theorems.

Theorem 3.2.1. Let P be a phrase of fuzzy literals from the set $\{x_i\}_{i=1}^n$. *A disjunction $\hat{\alpha}$ of any variable x_k and its complement $\bar{x}_k$, $1 \leqslant k \leqslant n$, can be appended to P without affecting the general value of the phrase iff there exists a variable x_i and its complement $\bar{x}_i$ in P for some i, $1 \leqslant i \leqslant n$.*

Proof.

(a) Assume x_i and $\bar{x}_i$ in P ($P = \alpha x_i \beta \bar{x}_i \gamma$, where α, β, and γ are conjunctions of literals from the set $\{x_i\}_{i=1}^n$). Obviously, $x_i \bar{x}_i \leqslant 0.5$ and thus $P \leqslant 0.5$. However, $\hat{\alpha} \geqslant 0.5$ and therefore the "if" part is proved.

(b) Assume that $x_i \bar{x}_i$ for some i, $1 \leqslant i \leqslant n$, is not in P. Then assign grades of membership larger than 0.5 to all literals of P. Clearly, $P \geqslant 0.5$. Q.E.D.

Similarly, we can prove the dual theorem.

Theorem 3.2.2. Let C be a clause of fuzzy literals from the set $\{x_i\}_{i=1}^n$. *A conjunction $\hat{\beta}$ of any variable x_k and its complement $\bar{x}_k$, $1 \leqslant k \leqslant n$, can be appended to C without affecting the general value of the clause iff there exists a variable x_i and its complement $\bar{x}_i$ in C for some i, $1 \leqslant i \leqslant n$.*

In general if F is a conjunction of formulas it can take a value $\leqslant 0.5$ if certain conditions are satisfied. Clearly if F is of the form $F = \left(\sum_j x_j \bar{x}_j \gamma_j \right) \beta$, $1 \leqslant j \leqslant n$, when β and $\{\gamma_j\}_{j=1}^n$ are formulas in $\{x_i\}_{i=1}^n$, then $F \leqslant 0.5$. This is a trivial case where one formula in the conjunction is $\leqslant 0.5$ and thus F is $\leqslant 0.5$. A more general case can be proved as follows.

Theorem 3.2.3. Let the set $\{F_i\}_{i=1}^\omega$ *be a set of fuzzy formulas over* $x_1, x_2, \ldots, x_n$, *and let F be a conjunction of formulas from this set. A disjunction F_d, of any formula F_k and its complement $\bar{F}_k$, can be appended to or deleted from the conjunction representing F without affecting the value of F, if there exist functions F_s and F_t in the conjunction representing F such that $\bar{F}_s$ is subsumed by F_t.*

Proof. Let F_s and F_t be in the conjunction representing F such that $\overline{F_s}$ is subsumed by F_t. Then $F_s F_t \leqslant 0.5$ since

$$\overline{F_s F_t} = \overline{F_s} + \overline{F_t} = \overline{F_s} + F_t + \overline{F_t} \geqslant 0.5. \qquad \text{Q.E.D.}$$

Theorem 3.2.4. Let the set $\{F_j\}_{j=1}^t$ be a set of fuzzy formulas over $x_1, x_2, \ldots, x_n$ and let F be a disjunction of formulas from this set. A conjunction F_c of any formula F_r and its complement $\overline{F_r}$ can be appended to or deleted from the disjunction representing F without affecting the value of F, if there exist functions F_i and F_l in the disjunction representing F such that F_i is subsumed by $\overline{F_l}$.

Proof. $F_i + F_l \geqslant 0.5$ since

$$F_i + F_l = F_i + \overline{F_l} + F_l. \qquad \text{Q.E.D.}$$

A fuzzy function can be specified by means of an arbitrary sum-of-products expression which is not a minimal sum and which can contain phrases which are not FPI's. This section presents a method to derive the set of all FPI's of a fuzzy function $f(x_1, x_2, \ldots, x_n)$ by means of the fuzzy iterative consensus.

Definition 3.2.3. Let R and Q be two phrases over the set of fuzzy variables $x_1, x_2, \ldots, x_n$. The *fuzzy consensus* of R and Q, written $R \, \psi \, Q$, is defined to be the set of phrases $\{R_i Q_i\}$, where $R = x_i R_i$ and $Q = \bar{x}_i Q_i$ (or $R = \bar{x}_i R_i$ and $Q = x_i Q_i$) and $x_i \in \{x_1, x_2, \ldots, x_n\}$, if the phrase $R_i Q_i$ includes the conjunction $x_j \bar{x}_j$ for at least one j, $j \in \{1, 2, \ldots, n\}$. If the phrase $R_i Q_i$ does not include $x_j \bar{x}_j$ for any j, $j \in \{1, 2, \ldots, n\}$, then

$$R \, \psi \, Q = \{R_i Q_i x_j \bar{x}_j \,|\, j = 1, 2, \ldots, n\} \quad x_i \in \{x_1, x_2, \ldots, x_n\}.$$

If none of the above occurs then we say that $R \, \psi \, Q = 0$. Any repeated literals or phrases are removed from the fuzzy consensus of R and Q.

Example 3.2.6.

(a) $R = x_1 x_2 \bar{x}_2 x_3, \qquad Q = \bar{x}_1 x_2 \bar{x}_2 x_3$
$R \, \psi \, Q = \{x_2 \bar{x}_2 x_3, \, x_1 \bar{x}_1 x_2 \bar{x}_2 x_3\}$

(b) $R = x_1 x_2, \qquad Q = x_1 \bar{x}_2$
$R \, \psi \, Q = \{x_1 \bar{x}_1, \, x_1 x_2 \bar{x}_2\}$

(c) $R = \bar{x}_1, \qquad Q = x_1 x_2 \bar{x}_2$
$R \, \psi \, Q = \{x_2 \bar{x}_2\}$

(d) $R = x_1\bar{x}_2 \qquad Q = \bar{x}_2 x_3$

$R \psi Q = 0.$

Theorem 3.2.5. Let R, Q, and W each represent a phrase. If $W \in R \psi Q$ then $R + Q \supseteq W$.

Proof. Since $W \in R \psi Q$ by definition W includes the conjunction $x_j \bar{x}_j$ for at least one j, $j \in \{1,2,\ldots,n\}$ or otherwise $W = 0$. If $W = 0$ clearly $R + Q \supseteq W$. If not so, four cases must be checked:

(1) $R \supseteq W$ and $Q \supseteq W$

(2) $R \supseteq W$ and $Q \not\supseteq W$

(3) $R \not\supseteq W$ and $Q \supseteq W$ $\qquad\qquad\qquad$ (3.2.16)

(4) $R \not\supseteq W$ and $Q \not\supseteq W$

In the first three cases it is clear that $R + Q \supseteq W$. We shall prove now the fourth case, namely $W \in R \psi Q$, $R \not\supseteq W$, and $Q \not\supseteq W$. Let R/W be the product of those literals of R which are not present in W and Q/W be the product of those literals of Q which are not present in W. For example, if $R = \bar{x}_1$, $Q = x_1 x_2 \bar{x}_2$ and $W = x_2 \bar{x}_2$, then $R/W = \bar{x}_1$ and $Q/W = x_1$. Then, since $W \in R \psi Q$, it must be true that

$R/W = (\overline{Q/W}),$

and since R/W and Q/W are both phrases, it follows that R/W must be a single literal, say x_i, and Q/W must be $\bar{x}_i$. Since $W \in R \psi Q$, by Theorem 3.2.1,

$$W = W(x_i + \bar{x}_i) = Wx_i + W\bar{x}_i. \qquad (3.2.17)$$

However,

$$R + Q \supseteq Wx_i \qquad (3.2.18)$$

and

$$R + Q \supseteq W\bar{x}_i \qquad (3.2.19)$$

and thus $R + Q \supseteq W$ since neither R nor Q can contain any literals other than x_i which are not present in W. $\qquad$ Q.E.D.

The phrases added whenever $x_j \bar{x}_j \not\in R_i Q_i$ are fuzzy prime implicants which are not needed. This can be seen from the following discussion.

Theorem 3.2.6. Let $R = x_i R_i$ and $Q = \bar{x}_i Q_i$ (or $R = \bar{x}_i R_i$ and $Q = x_i Q_i$) and $R_i Q_i$ does not include $x_j \bar{x}_j$ for any $j, j \in \{1,2,\ldots,n\}$. *Then*

$$R + Q + (R \psi Q) = R + Q$$

Proof.

$$
\begin{aligned}
R + Q + (R \psi Q) &= x_i R_i + \bar{x}_i Q_i + \{R_i Q_i x_j \bar{x}_j | j = 1,2,\ldots,n\} \\
&= x_i R_i + \bar{x}_i Q_i \\
&\quad + \{R_i Q_i x_j \bar{x}_j (x_i + \bar{x}_i) | j = 1,2,\ldots,n\} \\
&= x_i R_i + \bar{x}_i Q_i \\
&\quad + \{x_i R_i Q_i x_j \bar{x}_j + \bar{x}_i Q_i R_i x_j \bar{x}_j | j = 1,2,\ldots,n\} \\
&= x_i R_i + \bar{x}_i Q_i = R + Q.
\end{aligned}
$$

Q.E.D.

We shall define two kinds of fuzzy phrases. The first kind, to which we shall refer as type-1 phrase, are phrases which contain a conjunction of the form $x_j \bar{x}_j$ for at least one $j, j \in \{1,2,\ldots,n\}$. Otherwise we refer to the phrase as a type-2 phrase. Clearly a type-1 phrase cannot be subsumed by type-2 phrases. However, they can subsume some of them. For the case where members of the set $\{R_i Q_i\}$ do not include a conjunction of the form $x_j \bar{x}_j$ for at least one $j, j \in \{1,2,\ldots,n\}$, two situations must be checked.

(a) R and Q are both type-2 phrases. Since

$$\{R_i Q_i x_j \bar{x}_j | j = 1,2,\ldots,n\}$$

is a set of type-1 phrases covered by $R + Q$, this set is not needed.

(b) R is a type-1 phrase and Q is a type-2 phrase. In order for members of $\{R_i Q_i\}$ not to include a conjunction $x_j \bar{x}_j$ for any j, $j \in \{1,2,\ldots,n\}$, R must be of the form $\alpha x_i \bar{x}_i \beta$ and Q must be of the form $\gamma x_i \delta$ (or $\gamma' \bar{x}_i \delta'$) where the phrase $\alpha x_i \beta \gamma \delta$ (or $\alpha x_i \beta \gamma' \delta'$) is a type-2 phrase. Thus

$$R_i = \alpha x_i \beta \quad (\text{or } \alpha \bar{x}_i \beta)$$

$$Q_i = \gamma \delta \quad (\text{or } \gamma' \delta')$$

and obviously the set $\{R_i Q_i x_j \bar{x}_j | j = 1,2,\ldots,n\}$ is covered by Q and is not needed.

Example 3.2.7. Let $f(x_1, x_2, x_3) = x_1 x_2 + \bar{x}_2 x_3$. If the variables are

Boolean then the Boolean consensus will add the phrase x_1x_3. However, in the fuzzy case

$$x_j\bar{x}_j \not\subseteq x_1x_3, \; j \in \{1,2,3\}$$

and thus

$$x_1x_2 \; \psi \; \bar{x}_2x_3 = x_1x_3(x_1\bar{x}_1 + x_2\bar{x}_2 + x_3\bar{x}_3)$$

$$= x_1\bar{x}_1x_3 + x_1x_2\bar{x}_2x_3 + x_1x_3\bar{x}_3.$$

The phrase $x_1x_2\bar{x}_2x_3$ is not a fuzzy prime implicant since it subsumes both x_1x_2 and $\bar{x}_2x_3$. The phrases $x_1\bar{x}_1x_3$ and $x_1x_3\bar{x}_3$ are, however, fuzzy prime implicants, even though not essentials.

Theorem 3.2.7. Let

$$f(x_1,x_2,\ldots,x_n) = \prod_k \; (x_k + \bar{x}_k + \sigma_k)$$

where σ_k is an arbitrary function in $x_1,x_2,\ldots,x_n$ and $k \in \{1,2,\ldots,n\}$. Then

$$f(x_1,x_2,\ldots,x_n) = \prod_k \; (x_k + \bar{x}_k + \sigma_k) + \sum_j x_j\bar{x}_j\gamma_j$$

where γ_j is also some arbitrary function in $x_1,x_2,\ldots,x_n$ and $j \in \{1,2,\ldots,n\}$.

Proof.

$$\prod_k (x_k + \bar{x}_k + \sigma_k) = \text{Min} \left[\text{Max} \; (x_k,\bar{x}_k,\sigma_k) \right] \geqslant 0.5$$

$$\sum_j x_j\bar{x}_j\gamma_j = \text{Max} \left[\text{Min} \; (x_j,\bar{x}_j,\gamma_j) \right] \leqslant 0.5. \qquad\qquad \text{Q.E.D.}$$

It is interesting to note that the fuzzy consensus might add fuzzy prime implicants which include variables not even presented in the original fuzzy switching function. For example, let

$$f(x_1,x_2) = x_1 + \bar{x}_1.$$

Then

$$f(x_1,x_2) = x_1 + \bar{x}_1 + x_2\bar{x}_2$$

where all three phrases are fuzzy prime implicants. However,

$$x_2\bar{x}_2 = x_2\bar{x}_2(x_1 + \bar{x}_1) = x_2\bar{x}_2x_1 + x_2\bar{x}_2\bar{x}_1$$

and these two phrases are included in x_1 and $\bar{x}_1$, respectively. Hence

$$f_{\min}(x_1,x_2) = x_1 + \bar{x}_1$$

and $x_2\bar{x}_2$ is not an essential fuzzy prime implicant of the function.

3.3. MINIMIZATION OF COMPLETELY SPECIFIED FUZZY FUNCTIONS

The set of axioms of fuzzy algebra and Theorem 3.2.1 form the basis for the method of fuzzy iterated consensus and minimization of fuzzy switching functions. It will be shown in the following that the successive addition of fuzzy consensus phrases to a sum-of-products expression, and the removal of phrases which are included in other phrases $(x + xy = x)$ will result in an expression which represents the function as the sum of all its FPI's.

Theorem 3.3.1. A sum-of-products expression

$$F = P_1 + P_2 + \ldots + P_r$$

for the function $f(x_1,x_2,\ldots,x_n)$ is the sum of all the FPI's of $f(x_1,x_2, ,x_n)$ if and only if (1) no phrase includes any other phrase, $P_j \nsubseteq P_i$ for any i and j, $i \neq j$, $i,j \in \{1,2,\ldots,r\}$, and (2) the fuzzy consensus of any two phrases, $P_i \psi P_j$, either does not exist ($P_i \psi P_j = 0$3 or every phrase that belongs to the set describing $P_i \psi P_j$ is included in some other phrase from $\{P_k\}_{k=1}^{r}$.

Example 3.3.1. Let

$$f(x_1x_2) = \bar{x}_1 + x_1x_2\bar{x}_2, \quad \bar{x}_1 \psi x_1x_2\bar{x}_2 = x_2\bar{x}_2.$$

Adding $x_2\bar{x}_2$ to $f(x_1,x_2)$ implies that

$$f(x_1,x_2) = \bar{x}_1 + x_1x_2\bar{x}_2 + x_2\bar{x}_2,$$

and

$$f = \bar{x}_1 + x_2\bar{x}_2,$$

as $x_1x_2\bar{x}_2$ subsumes $x_2\bar{x}_2$ ($x_2\bar{x}_2$ includes $x_1x_2\bar{x}_2$).

Example 3.3.2. Let

$$f(x_1,x_2,x_3) = x_1x_2\bar{x}_2x_3 + \bar{x}_1x_2\bar{x}_2x_3$$

$$x_1x_2\bar{x}_2x_3 \; \psi \; \bar{x}_1x_2\bar{x}_2x_3 = \{x_2\bar{x}_2x_3, \; x_1\bar{x}_1x_2\bar{x}_2x_3\}.$$

Then

$$f(x_1,x_2,x_3) = x_1x_2\bar{x}_2x_3 + \bar{x}_1x_2\bar{x}_2x_3 + x_2\bar{x}_2x_3 + x_1\bar{x}_1x_2\bar{x}_2x_3,$$

and $f = x_2\bar{x}_2x_3$ as it includes all other phrases. The proof of Theorem 3.3.1 follows the principles presented in the proof of Boolean consensus.

Proof. It is necessary to show that the fact that

$F = P_1 + P_2 + \ldots + P_r$ is *not* the sum of all FPI's of $f(x_1,x_2,\ldots,x_n)$ implies that either $P_j \subseteq P_i$ for some i and j or that some phrase from the fuzzy consensus $P_i \; \psi \; P_j$ exists which is not included in some other phrase from the set $\{P_k\}_{k=1}^r$. It will now be assumed that

$$F = P_1 + P_2 + \ldots + P_r$$

is a sum-of-products expression but not the sum of *all* FPI's of

$$f(x_1,x_2,\ldots,x_n).$$

It will then be shown that there must be some P_i and P_j such that $P_j \subseteq P_i$ or that some phrase from the set describing $P_i \; \psi \; P_j$ exists for some i and j and there is no P_k that includes this phrase. There are two possible reasons for F not being the sum of all FPI's. Either some of the P_j's are not FPI's or some FPI's of $f(x_1,x_2,\ldots,x_n)$ are missing from F. If P_j is not an FPI, then there must be an FPI T of $f(x_1,x_2,\ldots,x_n)$ which includes P_j, $P_j \subseteq T$. If T occurs as one of the phrases, say $T = P_i$, then it follows that $P_j \subseteq P_i$. If T does not occur as one of the P_i's, then this is the situation where at least one of the FPI's is missing from F. It will thus be assumed next that there is some FPI T of $f(x_1,x_2,\ldots,x_n)$ which is missing from F. Since T is an FPI and is not identical with any of the P_i's it follows that $T \not\subseteq P_i$, $\forall i$, $i = 1,2,\ldots,r$.

Since T must be covered by the representation of the function, either $T \subseteq P_i$ for some i, or T is covered by a set of phrases, generated with respect to Theorem 3.2.3. However, $T \not\subseteq P_i$, $\forall i$, and therefore T must include at least one conjunction of the form $x_j\bar{x}_j$ for some j, $j \in \{1,2,\ldots,n\}$, in order for T to be covered in the representation of the function and to be an FPI of $f(x_1x_2,\ldots,x_n)$.

It may be possible to add some literals to T, forming a phrase T'

which still has the property that $T' \not\subseteq P_i$, $\forall i$. In general there may be several phrases satisfying the requirements placed on T'. In this case T' is defined as one of those phrases which satisfies the requirements given and contains as many literals as any other phrase satisfying these requirements. In case it is not possible to join an additional variable to T without having the resulting phrase included in one of the P_i's, T' is defined to be T itself.

In summary T' satisfies the following.

(1) T' contains no variables other than $x_1, x_2, \dots, x_n$.

(2) $T' \subseteq T$.

(3) $T' \not\subseteq P_i$, for any i, $i = 1, 2, \dots, r$.

(4) No phrase exists having more literals than T' and satisfying 1, 2, and 3 above.

It follows from the definition of T' that one of the variables will be missing from T'. If T' had all the set of variables appearing in it, condition (3) cannot be satisfied in the definition of T'.

Next the phrases formed by joining the missing variable, say x_k, and its complement, $\bar{x}_k$, to T', $T'x_k$, and $T'\bar{x}_k$ will be considered. Unless $T'x_k$ and $T'\bar{x}_k$ fail to satisfy condition (3), it is clear that T' could not have satisfied condition (4) as originally assumed. Thus $T'x_k$ and $T'\bar{x}_k$ must violate condition (3), and there must be some P_l and P_s such that $T'x_k \subseteq P_l$ and $T'\bar{x}_k \subseteq P_s$. From the fact that $T'x_k \subseteq P_l$ and $T' \not\subseteq P_l$, it follows that the literal x_k must appear in P_l so that it must be possible to express P_l as $P_l = x_k Q_l$. Similarly, it must be possible to express P_s as $P_s = \bar{x}_k Q_s$. It follows from the fact that $T'x_k \subseteq x_k Q_l = P_l$ that $T' \subseteq Q_l$, since neither Q_l nor T' contains x_k, and similarly that $T' \subseteq Q_s$, so that $T' \subseteq Q_l Q_s$ and thus the phrase $Q_l Q_s$ is non-zero. Clearly, $P_l \psi P_s$ is not empty, since at least $Q_l Q_s$ belongs to the set describing $P_l \psi P_s$. This is due to the fact that $Q_l Q_s$ includes $x_j \bar{x}_j$ for some j, since $T' \subseteq Q_l Q_s$.

All that remains to be shown is that there exist no P_i such that $Q_l Q_s \subseteq P_i$. This is not possible, for if $Q_l Q_s \subseteq P_i$, then because $T' \subseteq Q_l Q_s$ it follows that $T' \subseteq P_i$, which contradicts (3). This argument can be repeated for every phrase in the set describing $P_l \psi P_s$. In order to complete the proof of the theorem it is necessary to show that if $F = P_1 + P_2 + \dots + P_r$ is the sum of all the FPI's of $f(x_1, x_2, \dots, x_n)$, then (1) no phrase includes any other phrase $P_j \not\subseteq P_i$ for any i and j, $i \neq j$, $i, j \in \{1, 2, \dots, r\}$, and (2) the fuzzy consensus of any two phrases, $P_i \psi P_j$ either does not exist ($P_i \psi P_j = 0$) or every phrase that belongs to the set describing $P_i \psi P_j$ is included in some other phrase from $\{P_k\}_{k=1}^r$.

Each of the P_i's is an FPI by definition, and it follows from the definition of an FPI that it is not included in any other FPI of the same function. If $P_i \psi P_j$ exists, it is a set of phrases which is included in $f(x_1,x_2,\ldots,x_n)$. By the definition of an FPI any phrase which is included in $f(x_1,x_2,\ldots,x_n)$ must be included in some FPI of $f(x_1,x_2,\ldots,x_n)$, and thus every phrase that belongs to the set describing $P_i \psi P_j$ must be included in some other phrase from $\{P_k\}_{k=1}^r$.   Q.E.D.

Evidently the theorem

$$xy + \bar{x}z + yz = xy + \bar{x}z \qquad (3.3.1)$$

which is the basic Boolean consensus, does not hold in fuzzy logic. (Just take, for example, $\mu(x) = 0.7$, $\mu(y) = 0.8$, and $\mu(z) = 0.9$.) However, the following theorem is true in fuzzy logic.

Theorem 3.3.2. Let x, y and z be fuzzy variables. Then

$$x\bar{x}y + x\bar{y}z = x\bar{x}y + x\bar{y}z + x\bar{x}z. \qquad (3.3.2)$$

Proof. Expand $x\bar{x}z$ into $x\bar{x}z(y + \bar{y}) = x\bar{x}yz + x\bar{x}\bar{y}z$. The phrases $x\bar{x}yz$ and $x\bar{x}\bar{y}z$ are included in the phrases $x\bar{x}y$ and $x\bar{y}z$, respectively.   Q.E.D.

In order to make use of Theorem 3.3.1 in forming the sum of all FPI's from a sum-of-products expression, it is necessary to convert the original expression into one in which every phrase generated by the fuzzy consensus of any two phrases is included in some other phrase, and in which no phrase includes any other. This is done with the following algorithm.

To use the algorithm express $f(x_1,x_2,\ldots,x_n)$ in disjunctive normal form.

Algorithm 3.3.1.
   Input: The set of phrases representing $f(x_1,x_2,\ldots,x_n)$
   Output: The set of FPI's of $f(x_1,x_2,\ldots,x_n)$

Step 1: Compare each phrase with every other phrase in the expression and remove any phrase which subsumes any other phrase.

Step 2: Add the fuzzy consensus of any two phrases to the expression, provided that the fuzzy consensus phrases are not included in some other phrases.

Step 3: Remove any phrases that subsume phrases added in Step 2.

Steps 2 and 3 are iteratively repeated and the process terminates when all possible consensus operations have been exhausted. The remaining phrases are all of the FPI's. It should be noted that a phrase of the form $x_j\bar{x}_jAB$ is always included in the set describing $x_j\bar{x}_jA \ \psi \ x_j\bar{x}_jB$. This phrase, termed *secondary consensus,* need not be formed since it is always included in both of the initiating phrases.

Example 3.3.3. Let

$$f(x_1,x_2,x_3,x_4) = x_1\bar{x}_1x_2\bar{x}_3 + x_1\bar{x}_1x_2\bar{x}_4 + x_1\bar{x}_1x_2x_3x_4.$$

Let

$$R_1 = x_1\bar{x}_1x_2\bar{x}_3, \quad R_2 = x_1\bar{x}_1x_2\bar{x}_4, \quad R_3 = x_1\bar{x}_1x_2x_3x_4.$$

$$R_1 \ \psi \ R_3 = \{a_1,a_2\}$$

where

$$a_1 = x_1\bar{x}_1x_2x_3\bar{x}_3x_4 \text{and} \quad a_2 = x_1\bar{x}_1x_2x_4.$$

$$R_2 \ \psi \ R_3 = \{b_1,b_2\}$$

where

$$b_1 = x_1\bar{x}_1x_2x_3x_4\bar{x}_4 \quad \text{and} \quad b_2 = x_1\bar{x}_1x_2x_3.$$

Phrases a_2 and b_2 include phrases a_1 and b_1, respectively. Adding $R_1 \ \psi \ R_3$ and $R_2 \ \psi \ R_3$ to $f(x_1,x_2,x_3,x_4)$ yields, after removal of subsuming phrases,

$$f(x_1,x_2,x_3,x_4) = x_1\bar{x}_1x_2\bar{x}_3 + x_1\bar{x}_1x_2\bar{x}_4 + x_1\bar{x}_1x_2x_4 + x_1\bar{x}_1x_2x_3.$$

Let

$$R_4 = x_1\bar{x}_1x_2\bar{x}_3, \quad R_5 = x_1\bar{x}_1x_2\bar{x}_4,$$

$$R_6 = x_1\bar{x}_1x_2x_4, \quad R_7 = x_1\bar{x}_1x_2x_3.$$

$$R_4 \ \psi \ R_7 = \{x_1\bar{x}_1x_2x_3\bar{x}_3, \ x_1\bar{x}_1x_2\}.$$

The phrase $x_1\bar{x}_1x_2x_3\bar{x}_3$ is included in the phrase $x_1\bar{x}_1x_2$. Adding $R_4 \ \psi \ R_7$ to the function and removal of subsuming phrases will give us

$$f(x_1,x_2,x_3,x_4) = x_1\bar{x}_1x_2.$$

Usually it is a minimal sum rather than the sum of all FPI's of a fuzzy function that is desired.

In a sum-of-products form each phrase corresponds to a gate and each literal to an input line. The ratio between the cost of a gate and the cost of an input line will depend on the type of the gate. However, in practice the cost of an additional gate will be several times that of an additional input line on a gate which already exists. On this basis the elimination of gates will be the primary objective of the minimization process, leading to our standard definition of a minimal expression (Definition 3.2.1).

Theorem 3.3.3. A minimal sum-of-products expression must consist of a sum of phrases representing FPI's.

Proof. Consider the sum-of-products expression representing a function $f(x_1,x_2,\ldots,x_n)$. Any phrase which is not an FPI is, by definition, included in some FPI. Therefore we can replace this phrase in the set by any FPI including it. Since the FPI is of higher dimension, its representation will have fewer literals than that of the phrase it replaces.   Q.E.D.

Once all FPI's have been found (using the fuzzy iterated consensus algorithm), the job of finding the best set of FPI's remains. Before we present an algorithm to find certain FPI's that can be eliminated from the representation of the function without changing the function, we should like to emphasize that the minimum canonical sum-of-products form of a fuzzy function is not *necessarily* the union of *all* its FPI's.

A phrase is called a *fundamental phrase* if it does not include a conjunction of the form $x_j \bar{x}_j$ for at least one j, $1 \leq j \leq n$, or if it includes such a conjunction the phrase contains as many variables (complemented, uncomplemented, or both) as the function.

Definition 3.3.1. An FPI that includes a fundamental phrase which is not included in any other FPI is called an *essential FPI* (EFPI) and must be included in the corresponding minimal sum. (Actually, an EFPI must be included in all irredundant sums.)

To find a minimal sum of FPI's we construct a fuzzy prime implicant table, in a way similar to that used to construct prime implicant tables in Boolean algebra. Each column in the table corresponds to a fundamental phrase of $f(x_1,x_2,\ldots,x_n)$. At the left of each row are listed the FPI's of $f(x_1,x_2,\ldots,x_n)$.

An asterisk is placed at the intersection of a row and column if the corresponding FPI includes the corresponding fundamental phrase. In terms of the table the basic requirement on the minimal-sum phrases becomes that each column must have an asterisk in at least one of the rows which corresponds to minimal-sum phrases.

If any column contains only a single asterisk, the column corresponds to a distinguished fundamental phrase (it is included in only one FPI) and the row in which the asterisk occurs corresponds to an EFPI. Rows and columns corresponding to EFPI's and distinguished fundamental phrases are called essential rows and distinguished columns, respectively. The essential rows are checked and the labels of the distinguished columns are checked. If all the columns have been covered by the checked rows, all the fundamental phrases are included in the EFPI's and the sum of the EFPI's is the minimal sum-of-products expression. In case not all of the fundamental phrases are covered by EFPI's, we shall produce a reduced table. In this table all essential rows and all distinguished columns are eliminated.

Once the table has been reduced the remainder of the covering problem is to find the least expensive set of FPI's that covers the remainder of the function. This is done by implementing the same methods of reduction used in the minimization of Boolean functions on the reduced table.

Example 3.3.4. Let

$$f(x_1, x_2, x_3, x) = x_1 \bar{x}_1 x_2 + x_1 \bar{x}_1 x_3 x_4 + x_1 \bar{x}_2 x_3 x_4. \tag{3.3.3}$$

It can easily be seen that all three phrases are FPI's.

Expanding into fundamental phrases we obtain

$$x_1 \bar{x}_1 x_2 \rightarrow \left\{ x_1 \bar{x}_1 x_2 x_3 x_4, \ x_1 \bar{x}_1 x_2 \bar{x}_3 x_4, \ x_1 \bar{x}_1 x_2 x_3 \bar{x}_4, \ x_1 \bar{x}_1 x_2 \bar{x}_3 \bar{x}_4 \right\},$$

$$x_1 \bar{x}_1 x_3 x_4 \rightarrow \left\{ x_1 \bar{x}_1 x_2 x_3 x_4, \ x_1 \bar{x}_1 \bar{x}_2 x_3 x_4 \right\}, \tag{3.3.4}$$

$$x_1 \bar{x}_2 x_3 x_4 \rightarrow \left\{ x_1 \bar{x}_2 x_3 x_4 \right\}.$$

From Fig. 3.2. it can be seen that the phrases $x_1 \bar{x}_1 x_2$ and $x_1 \bar{x}_2 x_3 x_4$ are EFPI's. In this example all the fundamental phrases are included in the EFPI's and $x_1 \bar{x}_1 x_2 + x_1 \bar{x}_2 x_3 x_4$ is the minimal sum-of-products expression for the function.

3.4. MINIMIZATION OF INCOMPLETELY SPECIFIED FUZZY FUNCTIONS

So far the fuzzy switching functions considered have been completely specified for every combination of the variables. There exist situations, however, where, while a function is to assume a grade of membership for some combinations, it may assume unspecified grade memberships for a number of combinations. Combinations for which the grade membership of the function is not specified are called *don't care combinations*.

Fundamental phrases FPI's	$x_1\bar{x}_1x_2x_3x_4$	$x_1\bar{x}_1x_2\bar{x}_3x_4$ ✔	$x_1\bar{x}_1x_2x_3\bar{x}_4$ ✔	$x_1\bar{x}_1x_2\bar{x}_3\bar{x}_4$ ✔	$x_1\bar{x}_1\bar{x}_2x_3x_4$	$x_1\dot{\bar{x}}_2x_3x_4$ ✔
$x_1\bar{x}_1x_2$	*	*	*	*		
$x_1\bar{x}_1x_3x_4$	*				*	
$x_1\bar{x}_2x_3x_4$					*	*

FIG. 3.2. FPI table for Example 3.3.4.

The formulation of the *incompletely* specified single-output minimization problem can be stated as follows.

Given a fuzzy function F^* in the variables $x_1, x_2, \ldots, x_n$, constructed as a union of a completely specified subfunction and a don't-care subfunction, find the function F in the same variables such that F is the minimal complexity form of F^*.

We shall now present a theorem which will show how the FPI's of an incompletely specified fuzzy function can be obtained.

Theorem 3.4.1. Let F^* be an incompletely specified fuzzy function in the variables $x_1, x_2, \ldots, x_n$, expressed in disjunctive normal form. A phrase P is an FPI of F^* iff P is an FPI of $F_s + F_\Phi$ where F^* is represented by a pair of fuzzy functions F_s and F_Φ; F_s and F_Φ are the completely specified and the don't-care subfunctions, respectively.

Proof. (i) Assume $P_1 \in \{\text{FPI of } F^*\}$ and $P_1 \notin \{\text{FPI of } F_s + F_\Phi\}$. Thus, there exists P_2 such that $P_2 \in \{\text{FPI of } F_s + F_\Phi\}$ and P_1 subsumes P_2. Since P_2 is an implicant of F^*, P_1 cannot be an FPI of F^*. This is a contradiction, and hence $P_1 \in \{\text{FPI of } F_s + F_\Phi\}$.

(ii) Similarly, assume $P_1 \in \{\text{FPI of } F_s + F_\Phi\}$ and $P_1 \notin \{\text{FPI of } F^*\}$. Thus, there exists P_2 such that $P_2 \in \{\text{FPI of } F^*\}$ and P_2 subsumes P_1. Since P_1 is an implicant of F^*, part (i) implies that, if $P_2 \in \{\text{FPI of } F^*\}$, then $P_2 \in \{\text{FPI of } F_s + F_\Phi\}$. Hence, $P_1, P_2 \in \{\text{FPI of } F_s + F_\Phi\}$ and P_2 subsumes P_1. This is a contradiction, and hence

$$P_1 \in \{\text{FPI of } F^*\}. \hspace{3cm} \text{Q.E.D.}$$

The important consequence of this theorem is that to obtain the FPI's of an *incompletely specified fuzzy* function it is sufficient to consider the FPI's of the *fuzzy* function described by $F_s + F_\Phi$. This is an extended result with respect to Boolean *incompletely* specified functions.

To determine a minimal sum the FPI's of the incompletely specified fuzzy function are found. To do this attention is paid to both the phrases of F_s and F_Φ. The phrases of F_Φ are regarded as phrases of F_s during the formation of the FPI's.

Evidently, it is not necessary to include all the fundamental phrases of F_Φ in the minimal representation of the function. Thus, we can find all the FPI's of $F_s + F_\Phi$ using the algorithm 3.3.1. However, in constructing the fuzzy prime implicant table each column in the table corresponds only to a fundamental phrase of F_s. Hence columns are used only for the disjuncts of F_s, not of F_Φ, namely, each column in the table corresponds to a fundamental phrase of F_s. At the left of each row are listed the FPI's of $F_s + F_\Phi$.

Once the table has been reduced, the remainder of the covering problem is to find the least expensive set of FPI's that covers the remainder of the function F_s. This is done by implementing the same methods of reduction used in the minimization of Boolean functions and completely specified fuzzy functions.

Example 3.4.1. Let

$$F_s = x_1\bar{x}_2 + x_3x_4 + x_1\bar{x}_1x_2 + x_1x_2x_4 + x_2\bar{x}_3x_4 \qquad (3.4.1)$$

and

$$F_\Phi = x_1 + \bar{x}_1\bar{x}_2 + x_3 + \bar{x}_3 + x_4 + \bar{x}_4 + x_3\bar{x}_4. \qquad (3.4.2)$$

By the previous procedure we find that the FPI's of $F_s + F_\Phi$ are

$$x_1, x_3, \bar{x}_3, x_4, \bar{x}_4, \bar{x}_1\bar{x}_2. \qquad (3.4.3)$$

As can be easily seen from Fig. 3.3

$$F = F_{\min}^* = x_1 + x_4. \qquad (3.4.4)$$

Clearly, according to the construction of the algorithm, the algorithm is complete.

3.5. FUZZY MAPS

A direct consequence from §3.3 is that if a fuzzy phrase f_j subsumes another fuzzy phrase f_i, f_j can be deleted. This deletion will be performed repeatedly for all phrases of F until no two phrases subsume each other. In this section the concept of a fuzzy map is presented in order to find the set of subsumed implicants of the functions.

Theorem 3.5.1. The maximum number of fuzzy implicants in a disjunctive normal-form representation of a fuzzy function F of n variables is $4^n + 1$.

Proof. Using the rules of fuzzy algebra, a fuzzy function F of n variables can be expressed in the form

$$F(x_1, x_2, \ldots, x_n) = \sum_{k \in \Gamma} \left(\prod_{j \in S(k)} x_j \right) \qquad (3.5.1)$$

where Γ and $S(k)$ are index sets and $\bar{x}_j = x_{j+n}$. To illustrate the meaning of the above expression, let

Fundamental phrases of F_s / FPI's	$x_1\bar{x}_2$	x_3x_4	$x_1\bar{x}_1x_2x_3x_4$	$x_1\bar{x}_1x_2\bar{x}_3x_4$	$x_1\bar{x}_1x_2x_3\bar{x}_4$	$x_1\bar{x}_1\bar{x}_2\bar{x}_3x_4$	$x_1x_2x_4$	$x_2\bar{x}_3x_4$
x_1	*		*	*	*	*	*	
x_3		*	*		*			
$\bar{x}_3$				*		*		*
x_4		*	*	*			*	*
$\bar{x}_4$					*	*		
$\bar{x}_1\bar{x}_2$								

FIG. 3.3.   FPI table for Example 3.4.1.

$$F(x_1,x_2,x_3) = x_1\bar{x}_1x_2 + x_1x_2\bar{x}_3 + x_1x_3\bar{x}_3 + x_1x_2x_3$$

$$= \prod_{j\in S(1)} x_j + \prod_{j\in S(2)} x_j + \prod_{j\in S(3)} x_j + \prod_{j\in S(4)} x_j \qquad (3.5.2)$$

$$= \sum_{k\in\Gamma} \left(\prod_{j\in S(k)} x_j\right)$$

where $n = 3$, $\Gamma = \{1,2,3,4\}$, and $S(1) = \{1,2,4\}$, $S(2) = \{1,2,6\}$, $S(3) = \{1,3,6\}$, $S(4) = \{1,2,3\}$. Thus, every implicant except the number 0 can be represented as a binary string of length $2n$, where 1 in the ith place represents the variable x_i in the ith place and 0 in the ith place means that the implicant is vacuous in x_i. In general, there are $\binom{2n}{i}$ possible implicants with i 1's in their representation for $i = 0$, $1,\ldots, 2n-1$, $2n$. The number of these implicants is therefore

$$\sum_{i=0}^{2n} \binom{2n}{i} = 2^{2n} = 4^n \qquad (3.5.3)$$

where the string of $2n$ 0's represents the implicant $f_i = 1$. The implicant representing the number 0 is not included in the set of 4^n implicants, since it does not have a representation as a binary string of length $2n$ in our model. However, the number 0 has been included in the set of implicants and thus the total number of implicants is $4^n + 1$. Q.E.D.

The power of the Karnaugh map used in Boolean logic lies in its utilization of the ability of the human mind to perceive complex patterns in pictorial representations of data. The Karnaugh map may be regarded either as a pictorial form of a truth table or as an extension of the Venn diagram.

The *fuzzy map* may be regarded as an extension of the Veitch diagram which generated the basis for the Karnaugh map. It pictorially describes the set of all fuzzy implicants which represent the fuzzy function f. We interpret the universal set as the set of all 4^n combinations of values of $2n$ variables, divide this set into 4^n equal areas, and then place 1's in the areas corresponding to those combinations for which a fuzzy implicant appears in the description of the fuzzy function $f(x_1,x_2,\ldots,x_n)$.

We start with the universal set represented by a single area (Fig. 3.4a), and divide it into four areas corresponding to input combinations of 1, x_1, $\bar{x}_1$, and $x_1\bar{x}_1$ (Fig. 3.4b). We then divide it into four again, corresponding to 1, x_2, $\bar{x}_2$, and $x_2\bar{x}_2$ (Fig. 3.4c). With this notation, the interpretation of OR as the union of sets and AND as the intersection

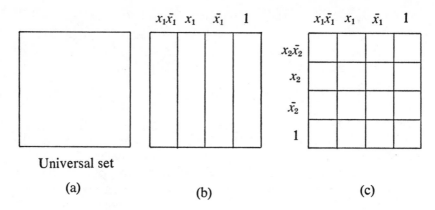

FIG. 3.4. Development of fuzzy maps.

of sets make it particularly simple to determine in which area we should place 1's (Fig. 3.5). Because of the commutative law we can change the order of the literals in each phrase without affecting the function. Thus the phrases will be represented as conjunctions of the form

$$x_1 x_{n+1} x_2 x_{n+2} \cdots x_j x_{j+n} \cdots x_n x_{2n} \qquad (3.5.4)$$

and *not* as

$$x_1 x_2 \cdots x_j \cdots x_n x_{n+1} \cdots x_{j+n} \cdots x_{2n-1} x_{2n}. \qquad (3.5.5)$$

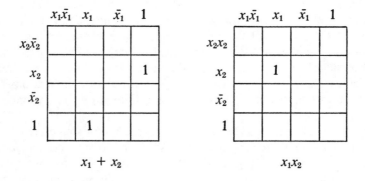

FIG. 3.5. OR and AND representations ($n = 2$).

We can replace the column and row headings with decimal numbers representing the binary headings. Thus, there are four combinations for each variable x_i, $1 \leq i \leq n$, to be presented in the heading of a row or a column.

(1) The heading is vacuous in x_i. The pair $x_i \bar{x}_i$ is denoted by 00 and represented by 0.

(2) The heading includes $\bar{x}_i$ but not x_i. The pair $x_i \bar{x}_i$ is denoted by 01 and represented by 1.

(3) The heading includes x_i but not $\bar{x}_i$. The pair $x_i \bar{x}_i$ is denoted by 10 and represented by 2.

(4) The heading includes x_i and $\bar{x}_i$. The pair $x_i \bar{x}_i$ is denoted by 11 and represented by 3.

Thus, every square on the fuzzy map is represented uniquely as a decimal number representing the binary sequence of length $2n$. The following illustration is given for a fuzzy map where $n = 2$, where the order is $x_1 \bar{x}_1 x_2 \bar{x}_2 (x_1 x_3 x_2 x_4)$.

The fuzzy map has a row dimension of 4^s and a column dimension of 4^{n-s} squares, where s and $n - s$ are the number of row and column variables, respectively.

	$x_1 \bar{x}_1$	x_1	$\bar{x}_1$	1
$x_2 \bar{x}_2$	15	11	7	3
x_2	14	10	6	2
$\bar{x}_2$	13	9	5	1
1	12	8	4	0

FIG. 3.6. Decimal representation of fuzzy implicants ($n = 2$).

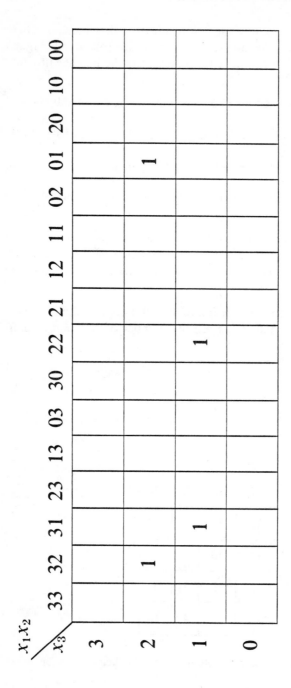

FIG. 3.7. The fuzzy map of $f(x_1), x_2, x_3) = \sum (6, 41, 53, 58)$

For simplicity the headings of the rows and columns will be expressed as decimal numbers. Thus, Fig. 3.6 will be represented as

	3	2	1	0
3	15	11	7	3
2	14	10	6	2
1	13	9	5	1
0	12	8	4	0

(with row header x_2 and column header x_1)

The fuzzy map can be considered as being a graphical representation of a table of combinations and hence a graphical representation of a fuzzy function. Figure 3.7 illustrates the fuzzy map of the following three-variables function:

$$f(x_1,x_2,x_3) = x_1\bar{x}_1x_2x_3 + x_1x_2\bar{x}_3 + \bar{x}_2x_3 + x_1\bar{x}_1\bar{x}_2\bar{x}_3. \tag{3.5.6}$$

Using our notation

$$f(x_1,x_2,x_3) = x_1x_2x_3x_4 + x_1x_2x_6 + x_5x_3 + x_1x_4x_5x_6 \tag{3.5.7}$$

$$= x_1x_4x_2x_3 + x_1x_2x_6 + x_5x_3 + x_1x_4x_5x_6$$

$$= \sum (111010,\ 101001,\ 000110,\ 110101)$$

$$= \sum (58,\ 41,\ 6,\ 53)$$

$$= \sum (6,\ 41,\ 53,\ 58).$$

The significance of the fuzzy map lies in the fact that it is possible to determine the implicants of a function from the patterns of 1's appearing on the map.

Define a square of the map with a 1 entry as being 1-square. The simplification procedure used on the fuzzy map is an exact translation of that used in fuzzy consensus theorem. However, this simplification procedure is quite different from the one used on a Karnaugh map. In Boolean logic two minterms can be "combined" by means of the theorem $xy + x\bar{y} = x$ if their corresponding binary representations differ in exactly 1 bit; namely, two minterms combine only if the corresponding points on an n-cube map are distance 1 apart, and thus they are adjacent squares on a Karnaugh map. On the fuzzy map, however, the simplification is performed through the subsuming operation, $x + xy = x$, and the fuzzy consensus result

$$\alpha \bar{x}_i x_i \beta + \bar{\alpha} x_i \bar{x}_i \beta = x_i \bar{x}_i \beta \qquad (3.5.8)$$

where $x_i \in \{x_1, x_2, \ldots, x_n\}$ and α and β are strings of literals from the set $\{x_j\}_{j=1}^n$. The technique for minimization is straightforward.

To restate the basic principles involved in the direction of a minimal complexity expression we state the following two steps: (1) choose as few groupings as possible, (2) choose each grouping as large as possible. Step (1) ensures that the number of first-level gates is reduced as much as possible, and step (2) ensures that each of these gates has a minimal number of input lines.

Theorem 3.5.2. An expression for a fuzzy function can be obtained by summing a set of phrases represented by rectangular groupings on a fuzzy map such that each subsumed 1-square is contained in at least one of the groupings.

Proof. By the fuzzy consensus operation

$$\alpha x_j \bar{x}_j \beta + \bar{\alpha} x_j \bar{x}_j \beta = x_j \bar{x}_j \beta$$

where $x_j \in \{x_1, x_2, \ldots, x_n\}$ and α and β are strings of literals from the set $\{x_i\}_{i=1}^n$. Thus, each rectangular grouping of 1-squares of dimension $2^k \times 2^q$, where $k \neq 0$ and $q \neq 0$ represents a phrase which includes a conjunction of the form $x_j \bar{x}_j$ for at least one j, $1 \leq j \leq n$. Clearly such a rectangular grouping is performed on a set of columns or rows which have the number 3 in the same digit of their headings. All other phrases are represented by rectangular groupings of dimension 1×1, namely $k = q = 0$. In case some rectangular groupings are subsumed the subsuming groupings are deleted from the map. The remaining groupings represent phrases which imply the fuzzy function, and thus it is sufficient to select the set of groupings such that every subsumed 1-square is included in at least one of these groupings. Hence, the function is properly described by an expression from the fuzzy map.   Q.E.D.

Definition 3.5.1. The right-bottom square on the fuzzy map, having 0's as row and column headings, is called the *dominant square*. This square is represented by the binary $2n$-tuple of 0's.

Definition 3.5.2. By a *fuzzy column extension* (FCE) we mean that any original square on the map in which a 1 is placed forms an extension of 1's throughout a set of squares G, where G is the set of squares which possesses a row heading which subsumes the original square row heading and which appears in the same column in which the original square appears.

Example 3.5.1. Let $f(x_1,x_2) = \bar{x}_1 + x_1\bar{x}_1x_2$.

	$x_1\bar{x}_1$	x_1	$\bar{x}_1$	1
$x_2\bar{x}_2$				
x_2	1			
$\bar{x}_2$				
1			1	

The fuzzy column extension of $f(x_1,x_2)$ is given by

	$x_1\bar{x}_1$	x_1	$\bar{x}_1$	1
$x_2\bar{x}_2$	1		1	
x_2	1		1	
$\bar{x}_2$			1	
1			1	

Similarly, a fuzzy row extension (FRE) can be defined. The reverse operations are called *fuzzy column condensation* (FCC) and *fuzzy row condensation* (FRC), respectively.

Theorem 3.5.3. The set of all needed FPI's of a fuzzy function can be obtained from the fuzzy map.

Proof. The proof is by construction. Let $f(x_1, x_2, \ldots, x_n)$ be a fuzzy function over n variables represented in disjunctive normal form. Given an n-variable fuzzy map, if there is a 1-entry on the dominant square, the function is identically equal to 1. This 1 is the only FPI of the function. If the dominant square does not have a 1-entry, perform all possible FCE's. Determine all rectangular groupings of 1-squares with maximal dimensions $(2^k \times 2^q = 2^{n-1})$ using the fuzzy consensus theorem

$$\alpha x_j \bar{x}_j \beta + \bar{\alpha} x_j \bar{x}_j \beta = x_j \bar{x}_j \beta \qquad (3.5.9)$$

Next, form all rectangular groupings of 1-squares with dimensions $2^k \times 2^q = 2^{n-j}$ for $j = 2$ such that no grouping is totally within a single previously formed grouping. Repeat this process for $j = 3, 4, \ldots, n$. It should be pointed out that since in Boolean logic all squares in the groupings on the Karnaugh map are 1-squares, they must describe prime implicants. However, on the fuzzy map one grouping may subsume another grouping, and thus not all the groupings are FPI's. In order to find the set of all FPI's perform all FRC's between complete rectangular groupings. Consider the implicant in row i and column j, where $1 \le i \le 4^s$, $1 \le j \le 4^{n-s}$. Every implicant in row i and in column k, such that column k possesses a heading that subsumes the heading of column j, will be deleted by the occurrence of 1 in row i and column j. This is due to the fact that entry ij on the map is subsumed by entry ik and is therefore deleted. However, this deletion can take place only on the basis of rectangular groupings; namely a complete grouping is deleted from the map if and only if it subsumes another complete grouping and not only part of it. The next step is to perform all possible FCC's. This operation is also performed on complete groupings only. In practice the above procedure is just an application of the fuzzy consensus theorem. Therefore the set of all phrases formed is the set of all needed F.P.I.'s of the function.   Q.E.D.

Algorithm 3.5.1.
Input: The set of phrases representing $f(x_1, x_2, \ldots, x_n)$.

Output: Minimal complexity form of $f(x_1, x_2, \ldots, x_n)$.

Step 1: Expand $f(x_1, x_2, \ldots, x_n)$ into a sum of fundamental phrases.
Step 2: Draw a fuzzy map of the expanded function.
Step 3: If there is a 1-entry on the dominant square, then $f = 1$ and go to step 8. Otherwise go to step 4.
Step 4: Perform all possible FCE's.

Step 5: Determine all possible groupings with respect to Theorem 3.5.1.

Step 6: Perform all possible FRC's. At this point a grouping will be deleted if it subsumes a set of 1-squares which do not necessarily have to be in a single grouping as required in Theorem 3.5.1.

Step 7: Perform all possible FCC's under the same conditions as step 6.

Step 8: Halt. The set of the remaining groupings represent the phrases of the minimal complexity form.

Example 3.5.2. Let

$$f(x_1,x_2) = x_1\bar{x}_2 + \bar{x}_1x_2 + x_1\bar{x}_1 + x_2\bar{x}_2. \tag{3.5.10}$$

The phrases $x_1\bar{x}_1$ and $x_2\bar{x}_2$ can be expressed as sum of fundamental phrases, and thus

$$f(x_1,x_2) = x_1\bar{x}_2 + \bar{x}_1x_2 + x_1\bar{x}_1x_2 + x_1\bar{x}_1\bar{x}_2 + x_1x_2\bar{x}_2 + \bar{x}_1x_2\bar{x}_2. \tag{3.5.11}$$

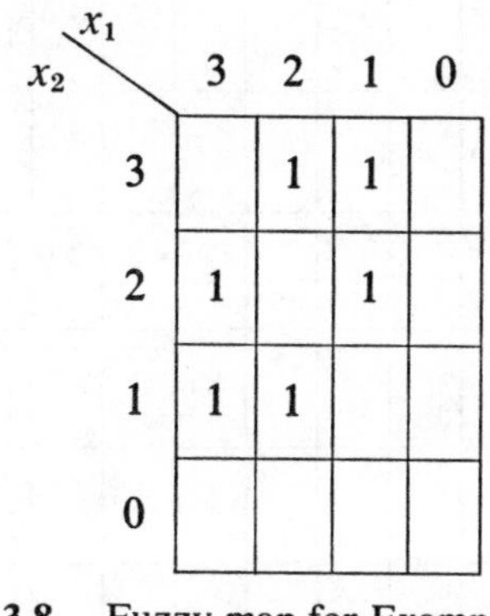

FIG. 3.8. Fuzzy map for Example 3.5.2.

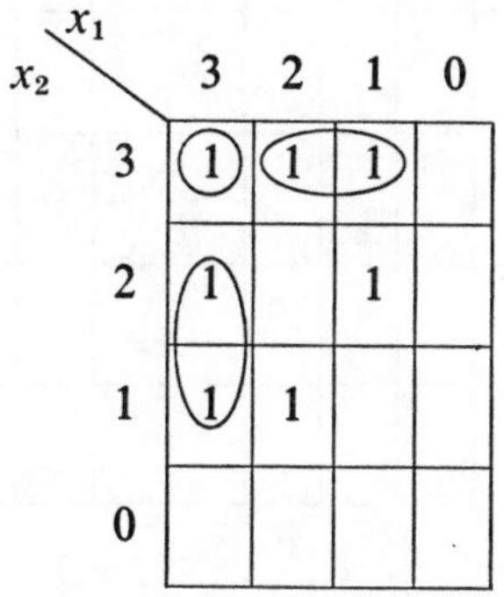

FIG. 3.9. Extended fuzzy map for Example 3.5.2.

x_3x_4 \ x_1x_2	33	32	31	23	13	03	30	22	21	12	11	02	01	20	10	00
33																
32				1	1											
31				1	1											
23																
13																
03																
30																
22		1	1						1							
21		1														
12		1														
11		1														
02											1					
01																
20																
10																
00																

FIG. 3.10. Fuzzy map for Example 3.5.3.

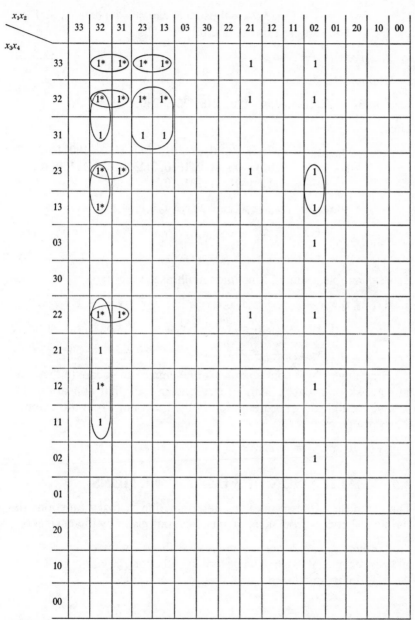

FIG. 3.11. Extended fuzzy map for Example 3.5.3. The circled terms are groupings of $2^k \times 2^q$ where k and q are not both 0. The terms marked with an asterisk are terms that can be deleted by FRC. Groupings in which all squares contain an asterisk are deleted.

The circled terms in Fig. 3.9 are those that can now be deleted by FRC and FCC. The rest are represented by

$$f_{\min}(x_1,x_2) = \bar{x}_1 x_2 + x_1 \bar{x}_2. \tag{3.5.12}$$

Example 3.5.3. Let

$$f = (34,\ 127,\ 128,\ 154,\ 190,\ 218,\ 229,\ 230,\ 233,\ 234).$$

Thus,

$$\begin{aligned}
f(x_1,x_2,x_3,x_4) = \sum\ (&00100010,\ 01111101,\ 01111110,\ 10011010,\\
&10111100,\ 11011010,\ 11100101,\ 11100110,\\
&11101001,\ 11101010)
\end{aligned}$$

$$\begin{aligned}
= &\ x_2 x_4 + \bar{x}_1 x_2 \bar{x}_2 x_3 \bar{x}_3 x_4 + \bar{x}_1 x_2 \bar{x}_2 x_3 \bar{x}_3 \bar{x}_4 + x_1 \bar{x}_2 x_3 x_4\\
&+ x_1 x_2 \bar{x}_2 x_3 \bar{x}_3 + x_1 \bar{x}_1 \bar{x}_2 x_3 x_4 + x_1 \bar{x}_1 x_2 \bar{x}_3 \bar{x}_4 + x_1 \bar{x}_1 x_2 \bar{x}_3 x_4\\
&+ x_1 \bar{x}_1 x_2 x_3 \bar{x}_4 + x_1 \bar{x}_1 x_2 x_3 x_4\ .
\end{aligned}$$

Expressing f as a sum of fundamental phrases we obtain

$$\begin{aligned}
f(x_1,x_2,x_3,x_4) = &\ x_2 x_4 + \bar{x}_1 x_2 \bar{x}_2 x_3 \bar{x}_3 x_4 + \bar{x}_1 x_2 \bar{x}_2 x_3 \bar{x}_3 \bar{x}_4 + x_1 \bar{x}_2 x_3 x_4\\
&+ x_1 x_2 \bar{x}_2 x_3 \bar{x}_3 x_4 + x_1 x_2 \bar{x}_2 x_3 \bar{x}_3 \bar{x}_4 + x_1 \bar{x}_1 \bar{x}_2 x_3 x_4 + x_1 \bar{x}_1 x_2 \bar{x}_3 \bar{x}_4\\
&+ x_1 \bar{x}_1 x_2 \bar{x}_3 x_4 + x_1 \bar{x}_1 x_2 x_3 \bar{x}_4 + x_1 \bar{x}_1 x_2 x_3 x_4.
\end{aligned}$$

Groupings which contain at least one unmarked square (either with an asterisk or a check mark) represent an FPI that appears in the minimal complexity representation of the function. From the union of these groupings:

$$f_{\min}(x_1,x_2,x_3,x_4) = x_1 \bar{x}_1 x_2 + x_2 \bar{x}_2 x_3 \bar{x}_3 + x_1 \bar{x}_2 x_3 x_4 + x_2 x_4.$$

3.6. REALIZATION OF FUZZY FUNCTIONS

The synthesis problem involves determination of fuzzy functions describing a process from a set of specifications given by the engineer.

Example 3.6.1. A certain process with attributes of interest $x,y,$ and z is recognizable when its describing function $f(x,y,z) \geq \alpha_1$. Obtain the describing function $f(x,y,z)$, if $f(x,y,z) \geq \alpha_1$ when

$$\left\{ \begin{array}{c} x \geq \alpha_1 \\ \text{and } y \leq 1 - \alpha_1 \end{array} \right\} \quad \text{or} \quad \left\{ \begin{array}{c} x \geq \alpha_1 \\ \text{and } y \geq \alpha_1 \\ \text{and } z \leq 1 - \alpha_1 \end{array} \right\}$$

$$\text{or} \quad \left\{ \begin{array}{l} x \leqslant 1 - \alpha_1 \\ \text{and } y \leqslant 1 - \alpha_1 \\ \text{and } z \geqslant \alpha_1 \end{array} \right\} \qquad (3.6.1)$$

The required fuzzy function is easily obtained by utilizing the procedure described above in reverse. The function is

$$f(x,y,z) = x\bar{y} + xy\bar{z} + \bar{x}\bar{y}z. \qquad (3.6.2)$$

The configuration of the electronic implementation of this function is the same as if the fuzzy variables were two-valued logic variables. If we use the symbols $\odot$, $\oplus$, and $\ominus$ to represent the three circuits in Figure 3.1a, b, and c, the electronic implementation of the function of equation (3.6.2) is shown in Fig. 3.13.

Example 3.6.2. If $f(x,y,z) \geqslant \alpha_1$ when

$$\left\{ \begin{array}{l} x \geqslant \alpha_{11} \\ \text{and } y \leqslant \alpha_{12} \end{array} \right\} \quad \text{or} \quad \left\{ \begin{array}{l} x \leqslant \alpha_{12} \\ \text{and } y \geqslant \alpha_{13} \end{array} \right\}$$

$$\text{or} \quad \left\{ \begin{array}{l} x \geqslant \alpha_{11} \\ y \geqslant \alpha_{13} \\ \text{and } z \leqslant \alpha_{14} \end{array} \right\} \qquad (3.6.3)$$

find the expression $f(x,y,z)$.

The essential difference between Example 3.6.2 and this one lies in the fact that the ranges of interest for x,y, and z are not necessarily related to the critical value α_1. Instead, arbitrary values such as α_{11}, α_{12}, α_{13}, etc. were selected. This, however, does not pose any problems if we keep in mind that an inequality of the type $x \geqslant \alpha_{1N}$ will depend on α_1 for its implementation, while one of the form $x \leqslant \alpha_{1M}$ will be dependent on $1 - \alpha_1$. This then implies that α_{1N} and α_{1M} must be properly multiplied (amplified) to attain the levels α_1 and $1 - \alpha_1$, respectively, prior to using them in an actual implementation. The required function is

$$f(x,y,z) = (k_{11}x) \cdot (k_{12}\bar{y}) + (k_{12}\bar{x}) \cdot (k_{13}y)$$

$$+ (k_{11}x) \cdot (k_{13}y) \cdot (k_{14}\bar{z}) \qquad (3.6.4)$$

where k_{mn} represents the multiplication factor to be associated with the various fuzzy variables. More specifically,

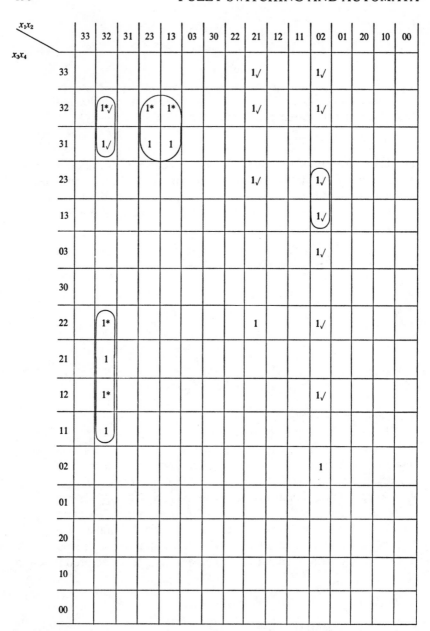

FIG. 3.12. Column condensation map for Example 3.5.3. The checked terms can be deleted by FCC. Groupings in which all squares contain an asterisk or a check (or both) are deleted.

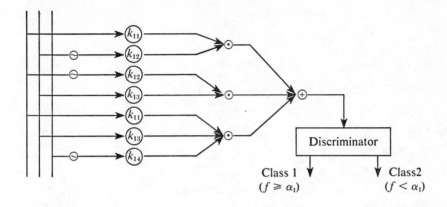

FIG. 3.13.   The realization of the fuzzy function of equation (3.6.2).

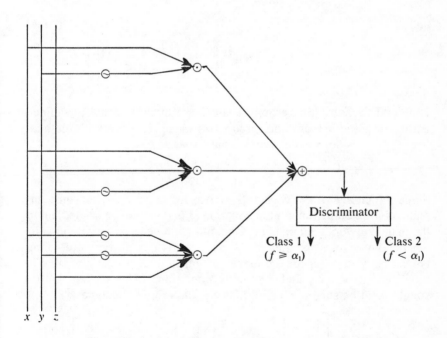

FIG. 3.14.   The realization of the fuzzy function of equation (3.6.4).

k_{11} in $k_{11}x$ is equal to α_1/α_{11}

k_{12} in $k_{12}\bar{y}$ is equal to $(1 - \alpha_1)/\alpha_{12}$

k_{12} in $k_{12}\bar{x}$ is equal to $(1 - \alpha_1)/\alpha_{12}$ (3.6.5)

k_{13} in $k_{13}y$ is equal to α_1/α_{13}

k_{14} in $k_{14}\bar{z}$ is equal to $(1 - \alpha_1)/\alpha_{14}$.

Figure 3.14 shows the electronic implementation of the fuzzy function of equation (3.6.4).

Example 3.6.3. Synthesize and implement electronically the fuzzy function $f(x,y,z)$ so that $a_M \leq f(x,y,z) < a_{M-1}$ when the fuzzy variables x, y, and z satisfy the conditions

$$\text{Group 1} = \left[\left\{ \begin{matrix} x \leq \alpha_1 \\ \text{and } y \geq \alpha_2 \end{matrix} \right\} \text{ or } \left\{ \begin{matrix} x \geq \alpha_3 \\ \text{and } y \geq \alpha_4 \\ \text{and } z \leq \alpha_5 \end{matrix} \right\} \right]$$

and   (3.6.6)

$$\text{Group 2} = \left[\left\{ \begin{matrix} x > \alpha_6 \\ \text{or } y < \alpha_7 \end{matrix} \right\} \text{ and } \left\{ \begin{matrix} x < \alpha_8 \\ \text{or } y < \alpha_9 \\ \text{or } z > \alpha_{10} \end{matrix} \right\} \right].$$

It should be noted that groups 1 and 2 are dual in nature as defined earlier. Making use of either group 1 or group 2, along with the analytical rules established above, we have that

$$f(x,y,z) = \bar{x}y + xy\bar{z} \tag{3.6.7}$$

Since the choice of the constants α_k was arbitrary, the electronic implementation of the function of equation (3.6.7) cannot be carried out in the usual conventional manner. One must first scale properly the various fuzzy variables and then develop the terms $x'y$ and xyz'. Prescaling is necessary in order for the fuzzy system to offer a response, that is $a_M \leq f(x,y,z) < a_{M-1}$ under suitable conditions. The scaling factors sought must be such that conditions equation (3.6.6) become

$$\text{Group 1} = \left[\left\{ \begin{matrix} x \leq 1 - a_M \\ \text{and } y \geq a_M \end{matrix} \right\} \text{ or } \left\{ \begin{matrix} x \geq a_M \\ \text{and } y \geq a_M \\ \text{and } z \leq 1 - a_M \end{matrix} \right\} \right]$$

and $(3.6.8)$

$$\text{Group 2} = \left[\left\{\begin{array}{c} x > 1 - a_{M-1} \\ \text{or} \quad y < a_{M-1} \end{array}\right\} \text{ and } \left\{\begin{array}{c} x < a_{M-1} \\ \text{or} \quad y < a_{M-1} \\ \text{or} \quad z > 1 - a_{M-1} \end{array}\right\}\right]$$

Thus, the scaling factors for the variables in groups 1 and 2 are as follows.

Group 1:

$$\alpha_1 k_{11} = 1 - a_m \quad \text{or} \quad k_{11} = (1 - a_m)/\alpha_1$$
$$\alpha_2 k_{12} = a_m \quad \text{or} \quad k_{12} = a_m/\alpha_2$$
$$\alpha_3 k_{13} = a_m \quad \text{or} \quad k_{13} = a_m/\alpha_3$$
$$\alpha_4 k_{14} = a_m \quad \text{or} \quad k_{14} = a_m/\alpha_4$$
$$\alpha_5 k_{15} = 1 - a_m \quad \text{or} \quad k_{15} = (1 - a_m)/\alpha_5$$

Group 2: $(3.6.9)$

$$\alpha_6 k_{21} = 1 - a_{m-1} \quad \text{or} \quad k_{21} = (1 - a_{m-1})/\alpha_6$$
$$\alpha_7 k_{22} = a_{m-1} \quad \text{or} \quad k_{22} = a_{m-1}/\alpha_7$$
$$\alpha_8 k_{23} = a_{m-1} \quad \text{or} \quad k_{23} = a_{m-1}/\alpha_8$$
$$\alpha_9 k_{24} = a_{m-1} \quad \text{or} \quad k_{24} = a_{m-1}/\alpha_9$$
$$\alpha_{10} k_{25} = 1 - a_{m-1} \quad \text{or} \quad k_{25} = (1 - a_{m-1})/\alpha_{10}.$$

In electronic parlance the scale factors k_{ij} are known as analog multipliers. Each variable must be scaled by its respective scale factor prior to being operated upon logically. Thus, in forming the product $\bar{x}y$ as required by group 1 of equation (3.6.6) one must first multiply x and y by k_{11} and k_{12}, respectively, subsequently complement x, and then form the product $\bar{x}y$; and similarly for the term $xy\bar{z}$. A more detailed description of the electronic implementation of this system is given in Fig. 3.15. It is clear from Fig. 3.14 that an output is generated if and only if $a_M \leqslant f(x,y,z) < a_{M-1}$. Note that the two threshold elements T_M and T_{M-1} are adjusted so that they respond when $f(x,y,z) \geqslant a_M$ and $f(x,y,z) \geqslant a_{M-1}$, respectively, except in the case in which $a_{M-1} = 1$. In that case threshold element T_{M-1} is not needed and thus may be removed.

Class selection may be varied by varying the boundaries a_M and a_{M-1} or, equivalently, the settings on the threshold elements T_M and T_{M-1}. These changes may be carried out either manually or automatically through program control. It is also conceivable that one might outline a set of n specifications such as those given in equation (3.6.6), in which case the previous synthesis procedure must be repeated n times.

In the above example, groups 1 and 2 are dual in nature. Duality in this example simply implies that groups 1 and 2 are from the same logical function. However, it is not necessary that this be so. The conditions on the fuzzy variables expressed by group 1 may be satisfied by $f_1 \geq a_M$ and those of group 2 by $f_2 \leq a_{M-1}$, where f_1 and f_2 are different functions. The functions f_1 and f_2 in such a case would be the system-describing functions and their electronic implementation is analogous to that given in Fig. 3.14.

Example 3.6.4. Consider the following statement.

A fuzzy system S has a single output R. Determine its electronic implementation so that it generates ar output $a_M \leq R < a_{M-1}$ when

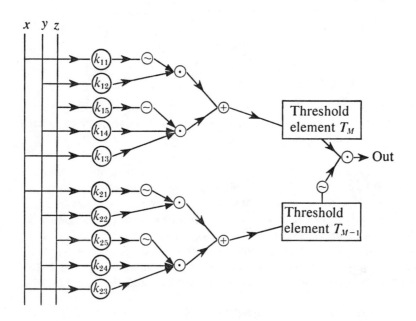

FIG. 3.15. The realization of equation (3.6.6).

$$\text{Group 1} = \left[\left\{\begin{matrix} & x \geqslant b_1 \\ \text{or} & y \leqslant b_2 \end{matrix}\right\} \text{ and } \left\{x \leqslant b_3\right\} \text{ and } \left\{z \geqslant b_4\right\}\right]$$

and   (3.6.10)

$$\text{Group 2} = \left[\left\{\begin{matrix} x \geqslant b_5 \\ \text{and } y \geqslant b_6 \end{matrix}\right\} \text{ or } \left\{z \leqslant b_7\right\}\right].$$

Groups 1 and 2 of equation (3.6.10) are not dual in nature and therefore each group must be associated with a different function. Let f_1 and f_2 be the two functions synthesized from group 1 and group 2, respectively. These functions are such that $f_1 \geqslant a_M$ and $f_2 \leqslant a_{M-1}$ when conditions equation (3.6.10) are satisfied.

Using the analytical rules stated previously, we can easily verify that

$$f_1(x,y,z) = (x + \bar{y}) \, \bar{x} \, z \qquad (3.6.11)$$

and

$$f_2(x,y,z) = (\bar{x} + \bar{y}) \, z. \qquad (3.6.12)$$

The scaling factors involved in this case are as follows.

Group 1	Group 2
$k_{11} = a_M/b_1$	$k_{21} = (1 - a_{M-1})/b_5$
$k_{12} = (1 - a_M)/b_2$	$k_{22} = (1 - a_{M-1})/b_6$
$k_{13} = (1 - a_M)/b_3$	$k_{23} = a_{M-1}/b_7.$
$k_{14} = a_M/b_4$	

(3.6.13)

The electronic implementation of this system is given in Fig. 3.16.

The circuits described in the first part of the section are not linear throughout the range of the variables and also are associated with the "deviation factor." The circuits described in the next part of this section have the following advantages: (1) higher fan-in and fan-out capabilities; (2) multiplication by a factor and the other conventional algebraic operations can be implemented in the gate itself; (3) they are linear throughout the range of the variables and the "deviation factor" is negligible.

The electronic implementation of a fuzzy OR function is shown in Fig. 3.17. Instead of using simple diodes active half-wave rectifiers are used so that the diode drop will not contribute too much to the "devia-

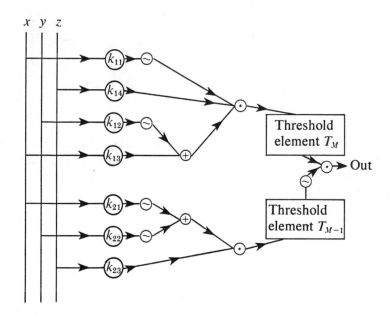

FIG. 3.16. The realization of equations (3.6.11) and (3.6.12).

tion factor." A half-wave rectifier circuit is shown in Fig. 3.19, where the fuzzy logic inverter is illustrated in Fig. 3.18.

When the input e_i is a positive quantity the output e_0' is also positive so the diode D_F conducts the switching resistance R_F into the feedback path. It can be shown that:

$$\frac{e_0'}{e_i} = \frac{R_G}{R_F' + R_G} \cdot \frac{R_F + R_G'}{R_G'} \tag{3.6.14}$$

where e_0' is the effective output and $R_F \gg r_f$, the forward resistance of the diode.

By choosing the values of the resistances the gain of the rectifier can be changed. In Fig. 3.17 the wired OR output of such rectifiers assumes the maximum value of the outputs of the rectifiers. The diode D_F in the rectifier circuit with maximum output will be conducting; other diode D_F's being reverse biased. Thus the output of the OR circuit is the maximum of the effective outputs of the rectifiers.

The circuit diagram of the inverter is shown in Fig. 3.19. It is functionally a difference amplifier. The output relation is as follows:

$$e_0 = (e_1 - e_2)k \tag{3.6.15}$$

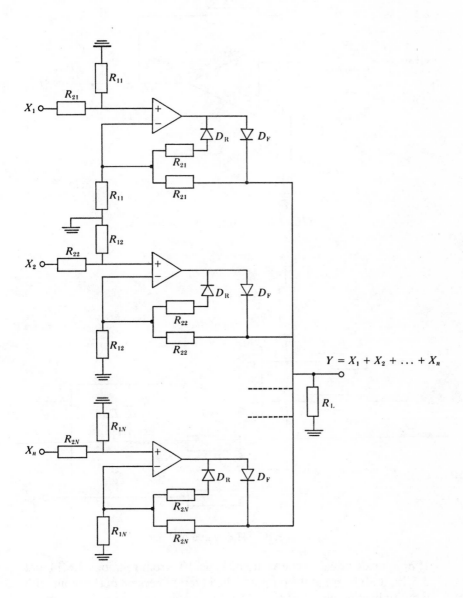

FIG. 3.17. Fuzzy OR gate.

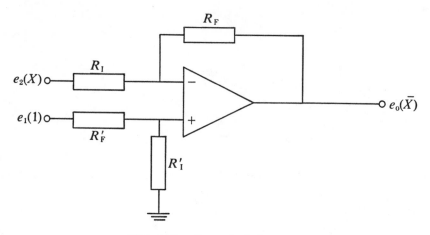

FIG. 3.18. Fuzzy logic inverter.

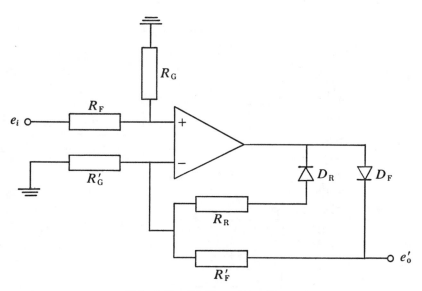

FIG. 3.19. Half-wave rectifier.

If e_1 is made equal to the value of logical 1, from relations 3.6.14 and 3.6.15, it is clear that the output is the inverted version of the input with a multiplication constant k.

The electronic implementation of the fuzzy AND function is shown in Fig. 3.20. To realize this circuit the following relation is used:

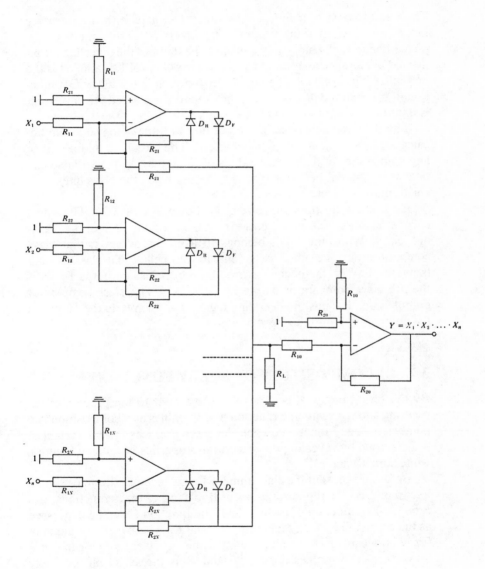

FIG. 3.20. Fuzzy AND gate.

$$x_1 \cdot x_2 \cdot \ldots \cdot x_n = \overline{(\bar{x}_1 + \bar{x}_2 + \ldots + \bar{x}_n)}. \qquad (3.6.16)$$

The circuit consists mainly of the fuzzy OR gate with the inverting inputs and the inverting output. The inversion of the input is implemented in each channel by changing the input to the inverting terminal and a reference level, equal to the value of logical 1, is applied at the non-inverting terminal. The wired OR function is the fuzzy OR function of the inverted inputs. Thus, the output of the wired OR, inverted, is the fuzzy AND function of the inputs.

The discriminator is an important unit in a fuzzy logic system because of its decision-making mechanism. The circuit should not have any ambiguities at the critical values of the input. As the variable can take any value in the range from 0 to 1, and when the class interval is small, more accurate circuits should be used.

An n-class discriminator circuit is shown in Fig. 3.21. The reference level of each class comparator is the maximum limit of the previous class. When the input belongs to a particular class, that class comparator and the other class comparators below that class assume the value 1. With the decoding circuit the pth-class output is 1 only if the pth-class comparator output is 1 and $(p + 1)$th-class comparator output is not 1. Thus the decoding logic circuit converts the output of the comparators into the 1 out of n code.

3.7. DECOMPOSITION OF FUZZY FUNCTIONS

By the term functional decomposition we refer to the process of expressing a fuzzy switching function of n variables as a composition of a number of fuzzy functions, each depending on less than n variables. This is done by revealing the structural properties of a fuzzy function while simplifying it.

Before we present the algorithm for fuzzy decomposition we should like to discuss the representations of Boolean and fuzzy functions. Let F be a fuzzy function of n variables. The function F can be expressed as the union of two fuzzy functions, F_n and F_f, where F_n is the union of fuzzy implicants in F which have the same structure as Boolean terms (however, the variables are fuzzy) and F_f is the set of all the fuzzy implicants of F which are not in F_n. The functions F_n and F_f are called the *normal fuzzy function* and the *free fuzzy function*, respectively. Members of F_n will be termed *fuzzy normal phrases* or *fuzzy minterms*.

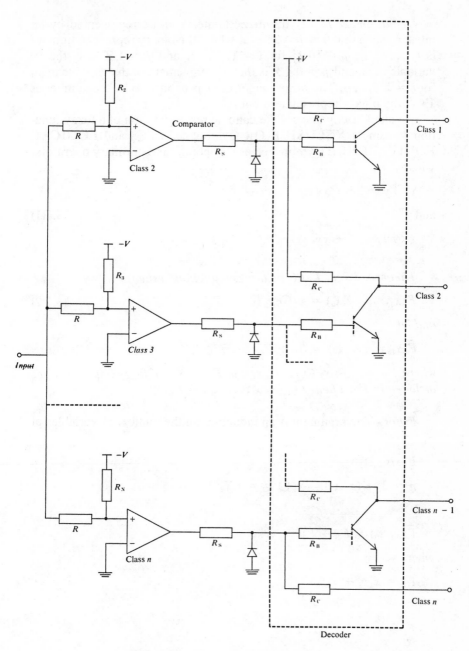

FIG. 3.21. An *n*-class discriminator.

Definition 3.7.1. Let the fuzzy minterm m_f be represented by an integer j such that $0 \leq j \leq 2^n - 1$, where the binary representation of j is $(j_1, \ldots, j_n)$, $j_i \in \{0,1\}$ for $i = 1, \ldots, n$, and $j_i = 1$ if x_i is the ith variable in m_f and $j_i = 0$ if $\bar{x}_i$ is the ith complement of the variable in m_f for $i = 1, \ldots, n$. The *fuzzy weight* of m_f is defined to be the number of 1's in the binary representation of m_f.

We shall briefly discuss the canonic representation of a fuzzy function F using EXCLUSIVE OR (symbolized as $\oplus$) and LOGICAL EQUIVALENCE (symbolized as $\odot$) operations as primary operators, where

$$x \oplus y = x\bar{y} + \bar{x}y$$

and   (3.7.1)

$$x \odot y = xy + \bar{x}\bar{y}.$$

Theorem 3.7.1. *Let F be a fuzzy function in the form*

$$F(x_1, x_2, \ldots, x_n) = x_1 \oplus x_2 \oplus \ldots \oplus x_n \qquad (3.7.2)$$

and let

$$\bar{F}(x_1, x_2, \ldots, x_n) = F_B + F_f,$$

where $F_B = x_1 \odot x_2 \odot \ldots \odot x_n$ and F_f the set of fuzzy implicants not included in F_B. Then, $\bar{F}_{min}(x_1, x_2, \ldots, x_n) = F_B$.

Proof. The argument is an induction on the number of variables of F.

Basis. If

$$n = 2, \quad F(x_1, x_2) = x_1 \oplus x_2 = x_1\bar{x}_2 + \bar{x}_1 x_2$$

and

$$\bar{F}(x_1, x_2) = F_B + F_f$$

where

$$F_B = x_1 x_2 + \bar{x}_1 \bar{x}_2$$

and

$$F_f = x_1\bar{x}_1 + x_2\bar{x}_2.$$

Then $x_1\bar{x}_1 + x_2\bar{x}_2$ can be expanded into

$$F_f = x_1\bar{x}_1(x_2 + \bar{x}_2) + x_2\bar{x}_2(x_1 + \bar{x}_1)$$

$$= x_1\bar{x}_1x_2 + x_1\bar{x}_1\bar{x}_2 + x_1x_2\bar{x}_2 + \bar{x}_1x_2\bar{x}_2.$$

Hence,

$$F = x_1x_2 + \bar{x}_1\bar{x}_2 + x_1\bar{x}_1x_2 + x_1\bar{x}_1\bar{x}_2 + x_1x_2\bar{x}_2 + \bar{x}_1x_2\bar{x}_2$$

and by removal of subsuming phrases

$$\bar{F} = x_1x_2 + \bar{x}_1\bar{x}_2. \tag{3.7.3}$$

Induction Step. Assume that the theorem is true for $n - 1$ variables.

$$F(x_1,x_2,\ldots,x_n) = x_1 \oplus x_2 \oplus \ldots \oplus x_{n-1} \oplus x_n$$

$$\bar{F}(x_1,x_2,\ldots,x_n) = \overline{x_1 \oplus \ldots \oplus x_n} = \overline{F(x_1,x_2,\ldots,x_{n-1}) \oplus x_n} \tag{3.7.4}$$

$$= x_nF(x_1,x_2,\ldots,x_{n-1}) + \bar{x}_n\bar{F}(x_1,x_2,\ldots,x_{n-1})$$
$$\quad + x_n\bar{x}_n + F(x_1,x_2,\ldots,x_{n-1})\,\bar{F}(x_1,x_2,\ldots,x_{n-1})$$

$$= x_nF(x_1,x_2,\ldots,x_{n-1}) + \bar{x}_n\bar{F}(x_1,x_2,\ldots,x_{n-1})$$
$$\quad + x_n\bar{x}_nF(x_1,x_2,\ldots,x_{n-1}) + x_n\bar{x}_n\bar{F}(x_1,x_2,\ldots,x_{-1})$$
$$\quad + x_nF(x_1,x_2,\ldots,x_{n-1})\,\bar{F}(x_1,x_2,\ldots,x_{n-1})$$
$$\quad + \bar{x}_nF(x_1,x_2,\ldots,x_{n-1})\,\bar{F}(x_1,x_2,\ldots,x_{n-1})$$

$$= x_nF(x_1,x_2,\ldots,x_{n-1}) + \bar{x}_n\bar{F}(x_1,x_2,\ldots,x_{n-1})$$

$$= \bar{F}(x_1,x_2,\ldots,x_{n-1}) \odot x_n.$$

However, by assumption

$$\bar{F}(x_1,x_2,\ldots,x_{n-1}) = x_1 \odot x_2 \odot \ldots \odot x_{n-1}$$

and thus

$$\bar{F}(x_1,x_2,\ldots,x_{n-1},x_n) = x_1 \odot x_2 \odot \ldots \odot x_{n-1} \odot x_n = F_B. \tag{3.7.5}$$
$$\text{Q.E.D.}$$

Similarly, we can prove the following theorem.

Theorem 3.7.2. Let F be a fuzzy function in the form

$$F(x_1,x_2,\ldots,x_n) = x_1 \odot x_2 \odot \ldots \odot x_n$$

and let

$$\bar{F}(x_1, x_2, \ldots, x_n) = F_C + F_g$$

where $F_C = x_1 \oplus x_2 \oplus \ldots \oplus x_n$ *and* F_g *is a set of fuzzy implicants not included in* F_C. *Then*

$$\bar{F}_{\min}(x_1, x_2, \ldots, x_n) = F_C.$$

In order to discuss the decomposition properties of fuzzy functions we introduce the following notation and terminology. Let F be a fuzzy function on a set X of n fuzzy variables, $x_1, x_2, \ldots, x_n$. A *partition* on a set of fuzzy variables X is a collection of non-empty, disjoint fuzzy subsets whose union is equal to X. Let $\{A, B\}$ be such a partition on X. If $f[\phi(A), B]$ takes on the same functional values as $F(X)$, whenever the latter are specified, then F is said to form a *fuzzy simple disjunctive decomposition* (FSD decomposition). The set of fuzzy variables A is the *bound fuzzy set,* and B is the *free fuzzy set.* Denote by m the number of bound fuzzy variables. A fuzzy function is said to be *fuzzily decomposable* if it has a non-trivial fuzzy simple disjunctive decomposition; thus $1 < m < n$. A fuzzy function which can be represented in the form of a fuzzy simple disjunctive decomposition has the multistage realization shown in Fig. 3.22.

When two or more FSD decompositions exist for a completely specified fuzzy function, they can be combined into a fuzzy complex disjunctive decomposition (FCD decomposition). Such an FCD decomposition has the form

$$F(X) = f\{\phi[\psi(A), B], \xi(C), D\} \qquad (3.7.6)$$

where F, f, ϕ, ψ, and ξ are fuzzy functions.

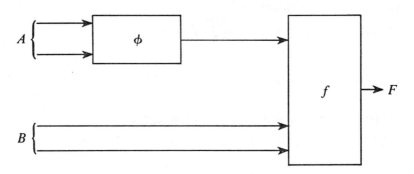

FIG. 3.22. FSD decomposition.

Because it is usually advantageous to realize two fuzzy functions of m and $n - m + 1$ variables rather than one fuzzy function of n variables (an immediate extension of the Boolean result), it is important to construct algorithms for finding all the FSD decompositions of a given fuzzy function. Curtis and Ashenhurst's method for determining decomposition of Boolean functions is based on the use of decomposition charts, which are rectangular arrays of decimal numbers corresponding to the binary input combinations. In the case of completely specified functions the decimal numbers are circled for 1's of the function and left blank for the 0's. Ashenhurst has proved the following theorem for identifying simple disjunctive decompositions of Boolean functions from a decomposition chart. In this theorem "column multiplicity" refers to the number of distinct Boolean vectors included in the set of columns.

Theorem 3.7.3. Let $\{A,B\}$ *be a partition on* X. *A completely specified switching function* $F(X)$ *(Boolean function) possesses a simple disjunctive decomposition with bound set* A *and free set* B *if and only if its* A/B *decomposition chart, with the variables in* A *defining the columns and variables in* B *defining the rows, has a column multiplicity* ρ *such that* $\rho \leq 2$.

Application of this theorem amounts to a straightforward inspection procedure for each A/B chart. There are 2^n possible charts, of which two are for partitions with one empty set and n are for partitions with a single bound variable ($n + 2$ trivial partitions). Furthermore, almost half of the charts can be eliminated by observing that the B/A chart is the A/B chart viewed sideways, and thus only $2^n - n - 2$ partitions must be checked on $2^{n-1} - 1$ charts.

Unfortunately, this basic theorem does not hold for fuzzy functions. First, let us show a counter example and then analyse the reasons that cause this theorem to fail in fuzzy logic.

Example 3.7.1. Let the fuzzy function F be

$$F(x_1,x_2,x_3,x_4) = \bar{x}_1\bar{x}_2\bar{x}_3x_4 + \bar{x}_1x_2\bar{x}_3x_4 + x_1\bar{x}_2\bar{x}_3x_4 + x_1x_2\bar{x}_3\bar{x}_4$$
$$+ \bar{x}_1\bar{x}_2x_3x_4 + \bar{x}_1x_2x_3x_4 + x_1\bar{x}_2x_3x_4.$$

The A/B decomposition chart of F corresponding to the Boolean representation of the variables and to the partition $A = \{x_1,x_2\}$, $B = \{x_3,x_4\}$ is the following.

x_3x_4 x_1x_2	00	01	10	11
00				1
01	1	1	1	
10				
11	1	1	1	

Therefore the *Boolean function* represented by F has a simple disjunctive decomposition of the form $F(\psi(x_1,x_2),x_3,x_4)$ where $\psi = x_1x_2$ and $f = \psi \, \bar{x}_3\bar{x}_4 + \bar{\psi}x_4$. However, if F is a fuzzy function (i.e. $x_j \in [0,1]$, $j = 1,2,3,4$ and not necessarily $x_j \in \{0,1\}$) it cannot be decomposed in a non-trivial way. The reason that fuzzy decompositions differ from Boolean decompositions is that in fuzzy algebra, unlike Boolean algebra, $x_i\bar{x}_i \neq 0$ and $x_i + \bar{x}_i \neq 1$, $i = 1,2,\dots,n$.

Following Shannon's expansion theorem about, say, x_1

$$F(x_1,x_2,\dots,x_n) = x_1 F(1,x_2,\dots,x_n) + \bar{x}_1 F(0,x_2,\dots,x_n)$$

where F is a Boolean function of n variables, we can easily verify the "fuzzy expansion form" for a fuzzy function F of n fuzzy variables achieved by successive applications of the algebraic rules given in § 0.0, where the fuzzy function F may be expanded about, say, x_1 as follows:

$$\begin{aligned} F(x_1,x_2,\dots,x_n) = {} & x_1 F_1(x_2,x_3,\dots,x_n) + \bar{x}_1 F_2(x_2,x_3,\dots,x_n) \\ & + x_1\bar{x}_1 F_3(x_2,x_3,\dots,x_n) + F_4(x_2,x_3,\dots,x_n) \end{aligned}$$

$$(3.7.7)$$

where $F_1,\dots,F_4$ are fuzzy functions of the $n - 1$ fuzzy variables, $x_2,\dots,x_n$. Thus

$$F(A,B) = f[\psi(A),B] = F(0,B) \, \bar{\psi}(A) + F(1,B) \, \psi(A)$$

is not true in general in fuzzy logic. Owing to this limitation it is clear that Ashenhurst's theorem for Boolean functions cannot be applied to fuzzy functions as it is, and different conditions must be presented.

First, we shall study the relations between the normal fuzzy function and Ashenhurst's theorem by means of the following theorem.

Theorem 3.7.4. Let $\{A,B\}$ be a partition on the fuzzy set X. A normal function is fuzzily decomposable with fuzzy bound set A and fuzzy free set B iff the fuzzy representation of the A/B decomposition chart, with the fuzzy variables in A defining the columns and the fuzzy variables in B defining the rows, has at most two distinct kinds of rows which can be classified into the following categories:

(a) (1) *a fixed pattern of 1's and 0's*
   (2) *all 0's*

or has at most four distinct kinds of rows which can be classified into the following categories:

(b) (1) *all 0's*
   (2) *all 1's*
   (3) *all fuzzy minterms of even weight in the row*
   (4) *all fuzzy minterms of odd weight in the row.*

Proof. The "if" part is quite obvious as we can write the normal fuzzy function in its fuzzily decomposed form by denoting the rows in the first category in (a) by ψ or denoting the third category in (b) by ψ and the fourth by $\bar{\psi}$, and the second in (b) by $\psi + \bar{\psi}$. To prove the "only if" part, we have to investigate the definitions of all logical connectives in fuzzy algebra. It is easy to see that complements of logical connectives are defined to fit into the scheme of Boolean decompositions because $x_i + \bar{x}_i = 1$ and $x_i \bar{x}_i = 0$, $\forall i$. Because this is not true in fuzzy albegra we have to exclude these logical connectives in fuzzy decomposition. Checking the set of logical connectives leaves us with EXCLUSIVE OR ($\oplus$) and its complement, the EQUIVA-LENCE ($\odot$), by Theorem 3.7.1 or its dual Theorem 3.7.2. If we construct a fuzzy representation of a decomposition chart as described in the theorem, in case (a) the fixed pattern of 1's and 0's corresponding to the case $F_n(A,B) = f[\psi(A),B]$ based on the normal fuzzy function expansion

$$F_n(x_1, x_2, \ldots, x_n) = x_j \, F_1 \, (x^j) + \bar{x}_j F_2(x^j)$$

where

$$x^j = (x_1, x_2, \ldots, x_{j-1}, x_{j+1}, \ldots, x_n).$$

Based on the same expansion we can check case (b) where

$$F_n(A,B) = f[\psi(A),B] = f_1(B) \, \psi \, (A) + f_2(B) \, \bar{\psi} \, (A).$$

The four categories in case (b) are due to the four relations between $f_1(B)$ and $f_2(B)$. However, these relations are controlled by the fact that $\overline{\psi}(A)$ is the complement of $\psi(A)$, and as shown above this restricts the representation to cases where $\psi(A)$ is the EXCLUSIVE OR relation of the fuzzy variables of A. Q.E.D.

Example 3.7.2. Let

$$F_n(x_1, x_2, x_3, x_4, x_5) = \sum(1, 3, 6, 7, 10, 11, 13, 15, 18, 19, 21, 23, 25,$$
$$27, 30, 31):$$

x_4x_5 \ $x_1x_2x_3$	000	001	010	011	100	101	110	111
00								
01	1			1		1	1	
10		1	1		1			1
11	1	1	1	1	1	1	1	1

This can be fuzzily decomposable by case (b):

$$F_n(x_1, x_2, x_3, x_4, x_5) = x_4\bar{x}_5(x_1 \oplus x_2 \oplus x_3) + \bar{x}_4x_5\overline{(x_1 \oplus x_2 \oplus x_3)}$$
$$+ x_4x_5\big[(x_1 \oplus x_2 \oplus x_3) + \overline{(x_1 \oplus x_2 \oplus x_3)}\big].$$

Thus,

$$F_n(x_1, x_2, x_3, x_4, x_5) = \xi(\psi, x_4, x_5)$$

where

$$\psi(x_1, x_2, x_3) = x_1 \oplus x_2 \oplus x_3$$

and

$$\xi(\psi, x_4, x_5) = x_4\bar{x}_5\,\psi + \bar{x}_4x_5\,\overline{\psi} + x_4x_5(\psi + \overline{\psi}).$$

The realization of F_n is given in Figure 3.23.

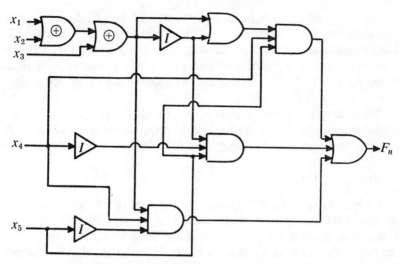

FIG. 3.23. Realization of Example 3.7.2.

3.8. FSD DECOMPOSITION

It should be emphasized that the fact that the normal fuzzy function F_n is fuzzily decomposable does not imply necessarily that the entire fuzzy function F is fuzzily decomposable and *vice versa*. This will be demonstrated by the following example.

Example 3.8.1. Let

$$F_n = x_1\bar{x}_2x_3\bar{x}_4 + x_1\bar{x}_2x_3x_4 + x_1x_2\bar{x}_3\bar{x}_4 + x_1x_2\bar{x}_3x_4 + x_1x_2x_3x_4.$$

The A/B decomposition chart of F_n corresponding to the *Boolean* representation of the variables and to the partition $A = \{x_1, x_2\}$ $B = \{x_3, x_4\}$ is the following:

x_3x_4 \ x_1x_2	00	01	10	11
00				1
01				1
10			1	
11			1	1

Owing to Theorem 3.7.4 the Boolean representation F_n does not have a simple disjunctive decomposition of the form $f[\psi(A),B]$. However, it will be shown later that there exist fuzzy functions such that F_n above is their normal fuzzy function, and even if F_n as shown here is not decomposable these fuzzy functions as a whole are fuzzily decomposable.

We shall now present an algorithm for FSD (Fuzzy Simple Disjunctive) decomposition of fuzzy functions. The algorithm is a result of the previous theorems and the fuzzy map representation described in § 3.5.

Algorithm 3.8.1.

Step 1: Given a fuzzy function F over a set of n fuzzy variables X, construct the fuzzy map of dimension $4^{n-s} \times 4^s$ with the fuzzy variables in A defining the columns and the fuzzy variables in B defining the rows when A/B is a partition on X. If the fuzzy map has at most three distinct kinds of rows which can be classified into the categories (1) all 0's (2) a fixed pattern of 1's and 0's, and (3) rows with a single 1 in the 1 column (phrases which include variables only from set B), then F is fuzzily decomposable and go to step 9; otherwise go to step 2.

Step 2: Express F as the union of F_n and F_f. Check if F_n is fuzzily decomposable by Theorem 3.7.4. If the answer is "yes" go to step 3; otherwise go to step 5.

Step 3: Check if every row in the submap of dimension $2^{n-s} \times 4^s$, where the 2^{n-s} columns are defined by the set A in F_n, has the form of the rows in the fuzzy representation of the decomposition chart of F_n. If "yes" go to step 4; otherwise go to step 5.

Step 4: Perform the minimization procedure on the $4^{n-s} \times 4^s$ map without affecting the fuzzy decomposition criteria established by Theorem 3.7.4 on the $2^{n-s} \times 4^s$ map discussed in step 3. (Thus deletion of complete rows on this table, or part of them, without changing the classification of rows by Theorem 3.7.4 is allowed.) If what is left over is the decomposed table $2^{n-s} \times 4^s$ and any pattern of 1's and 0's in the 1 column (the right-hand-side column) go to step 9; otherwise go to step 6.

Step 5: Perform the minimization procedure, one step at a time, and check after each step if the decomposition criteria for the $2^{n-s} \times 4^s$ map are fulfilled. If the answer is "yes" in at least one step, stop the minimizatization and go to step 4; otherwise go to step 10.

Step 6: If the $2^{n-s} \times 4^s$ map is empty, go to step 7; otherwise go to step 10.

Step 7: Check the categories in step 1. If they check, go to step 9; otherwise go to step 8.

Step 8: Check if the fuzzy map has at most three distinct kinds of columns which can be classified into the following categories:

(a) (1) all 0's
    (2) a pattern of 1's and 0's in the 1 column
    (3) an arbitrary pattern of 1's and 0's in any other single column

or has at most four distinct kinds of columns which can be classified into the following categories:

(b) (1) all 0's
    (2) a pattern of 1's and 0's in the 1 column
    (3) two arbitrary patterns of 1's and 0's in columns possessing the property that their headings fulfil the DeMorgan laws.

If "yes" in either (a) or (b) go to step 9; otherwise go to step 10.

Step 9: Terminate the algorithm. The fuzzy function F is fuzzily decomposable.

Step 10: Terminate the algorithm. The fuzzy function F is not fuzzily decomposable by the criteria specified in this section. Breaking the algorithm into a set of lemmas and their simple proofs will be omitted here.

Example 3.8.2. Let

$$F = x_1\bar{x}_1x_2\bar{x}_2\bar{x}_3\bar{x}_4 + x_1\bar{x}_1\bar{x}_2x_3x_4 + x_1\bar{x}_2x_2x_3x_4 + \bar{x}_1x_2x_3x_4 + x_1\bar{x}_2x_3x_4$$
$$+ x_1x_2\bar{x}_3x_4 + \bar{x}_1\bar{x}_2\bar{x}_3x_4 + x_1x_2\bar{x}_3\bar{x}_4 + \bar{x}_1x_2\bar{x}_3\bar{x}_4 + x_1\bar{x}_2\bar{x}_3\bar{x}_4$$
$$+ \bar{x}_1\bar{x}_2\bar{x}_3\bar{x}_4 + x_4\bar{x}_4 + x_3x_3 + x_3\bar{x}_4$$

The fuzzy map of F corresponding to the partition $A = \{x_1, x_2\}$, $B = \{x_3, x_4\}$ is described in Fig. 3.24. Therefore, F has a fuzzy simple disjunctive decomposition of the form

$$F(x_1, x_2, x_3, x_4) = f[\psi(x_1, x_2), x_3, x_4]$$

where

$$\psi(x_1, x_2) = \bar{x}_1x_2 + x_1\bar{x}_2$$

and

$$f = \psi(x_1, x_2)(x_3x_4 + \bar{x}_3\bar{x}_4) + \bar{\psi}(x_1, x_2)(\bar{x}_3x_4 + \bar{x}_3\bar{x}_4)$$
$$+ x_3\bar{x}_4 + x_3\bar{x}_3 + x_4\bar{x}_4.$$

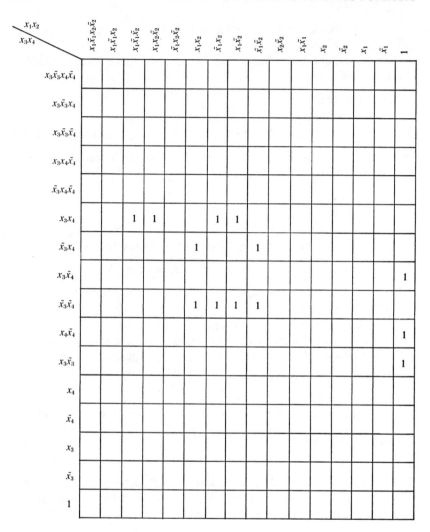

FIG. 3.24. Fuzzy map for Example 3.8.2.

Example 3.8.3. Let

$$F = x_1\bar{x}_2x_3\bar{x}_4 + x_1\bar{x}_2x_3x_4 + x_1x_2\bar{x}_3\bar{x}_4 + x_1x_2x_3x_4 + x_1x_2\bar{x}_3x_4$$
$$+ \bar{x}_3x_4\bar{x}_4 + x_3.$$

This function has the same F_n as the function in Example 3.8.1. Even though the Boolean representation of F_n does not have a simple disjunctive decomposition of the form $f[\psi(A), B]$, the fuzzy function F has an FSD decomposition of the form

$$F = f\big[\theta(x_1,x_2),\, x_3,\, x_4\big]$$

where

$$\theta(x_1,x_2) = x_1 x_2$$

and

$$f = \theta(x_1,x_2)\,(\bar{x}_3 x_4 + \bar{x}_3 \bar{x}_4) + \bar{x}_3 x_4 \bar{x}_4 + x_3.$$

3.9. ANOTHER VIEW OF DECOMPOSITION

The importance of the graphical representation of Boolean switching function is clear to anyone who has used a Karnaugh map or a Vietch diagram. The advantage of such a pictorial representation of logic functions lies in the relative ease with which relational patterns among the implicants of a function can be perceived. It is the purpose of this section to present a scheme for graphical presentation and manipulation of fuzzy switching functions. The use of a fuzzy map for such manipulation gives insight into the structure of the function and relations between its implicants. This insight can be of great value in the understanding and solution of problems such as minimization and decomposition of such functions.

It has been asserted that, owing to the continuously valued nature of fuzzy variables, graphical representation in a manner analogous to that used in Boolean logic is not possible. This misapprehension is perhaps due to misunderstanding of the function of the symbols 1 and 0 often written within the boundaries of a Boolean logic map. These symbols do not denote the values of terms, only their presence or absence in the function. Accordingly, since there can be only a finite number of (unique) terms in a fuzzy switching function of finitely many variables, we can represent any term or set of terms as subsets of a logical space.

A map representation of two-variable fuzzy switching functions (hereinafter called FSF's) was introduced in § 3.5. The following will introduce and re-emphasize some basic properties of this map. Figure 3.26 shows a map in which a logical space is subdivided into 16 squares, each corresponding to one of the 16 non-zero implicant

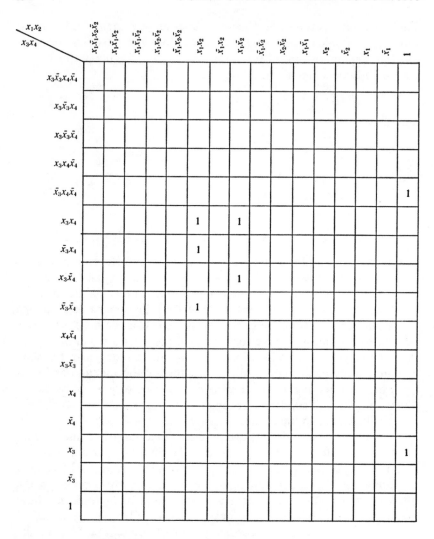

FIG. 3.25. Fuzzy map for Example 3.8.3.

phrases of a disjunctive normal form FSF of two variables. The presence of the symbol I in a square denotes the presence of the associated implicant in the mapped function.

This map representation has important properties which distinguish it from a Boolean map. Perhaps the most obvious of these is that the squares (each representing a different implicant) are not disjoint in the

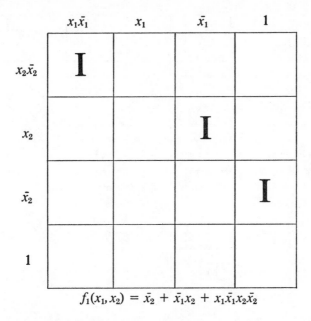

$$f_1(x_1, x_2) = \bar{x}_2 + \bar{x}_1 x_2 + x_1 \bar{x}_1 x_2 \bar{x}_2$$

FIG. 3.26.   A two-variable fuzzy map.

sense that the absence of an I in a square does not necessarily exclude the associated implicant from an equivalent representation of the function. Conversely, the presence of an I does not imply essentiality of the associated implicant. In the example of Fig. 3.26 the implicant $x_1 \bar{x}_2$ could be added to the function $f_1(x_1, x_2)$ without changing the function. Similarly, the deletion of the implicant $x_1 \bar{x}_1 x_2 \bar{x}_2$ will not alter f_1. These facts are a direct consequence of the use of Min and Max operators in the fuzzy algebra. One can observe (Fig. 3.27) distinct patterns of included (subsuming) implicants associated with each implicant square on the two-variable map. It is clear that recognition of such patterns is crucial to minimization of FSF's, as will be illustrated in a later section.

Since each fuzzy implicant includes some number of other implicants, we define the *cover fraction* (CF) of a square on the map as the fraction of the map included (covered) by the implicant associated with that square. Figure 3.28 displays the cover fraction of each square on the two-variable map in decimal ratios and as hexadecimal fractions. The hex notation is more natural and will be used in the rest of this paper. The cover fraction concept may be used to indicate the pattern

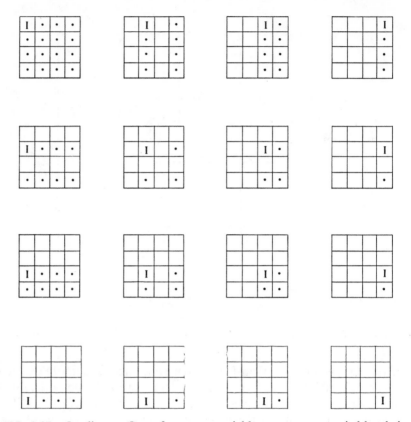

FIG. 3.27. Implicants (I) on fuzzy two-variable maps accompanied by their retinues of included (subsuming) implicant (·).

included by any implicant square I. We first define two sets of columns and rows with the following properties: for any implicant square I, let

$A \equiv \{$columns$|$I's column $\cup$ any column containing a lower-CF square in I's row.$\}$

$B \equiv \{$rows$|$I's row $\cup$ any row containing a lower-CF square in I's column.$\}$

The pattern associated with I is then the Cartesian product $A \times B$. In each case I is the largest-CF member of $A \times B$. The existence of included but distinct implicants of the fuzzy map contrasts with the impossibility of such inclusion on a Karnaugh map (barring consideration of blocks of neighboring Boolean implicant-squares "including" their subsets).

$\frac{1}{16}$	$\frac{2}{16}$	$\frac{2}{16}$	$\frac{4}{16}$
$\frac{2}{16}$	$\frac{4}{16}$	$\frac{4}{16}$	$\frac{8}{16}$
$\frac{2}{16}$	$\frac{4}{16}$	$\frac{4}{16}$	$\frac{8}{16}$
$\frac{4}{16}$	$\frac{8}{16}$	$\frac{8}{16}$	$\frac{16}{16}$

0.1	0.2	0.2	0.4
0.2	0.4	0.4	0.8
0.2	0.4	0.4	0.8
0.4	0.8	0.8	1.0

(a)   (b)

FIG. 3.28. Cover function of two-variable map implicant squares expressed as (a) decimal ratios and (b) hexadecimal fractions.

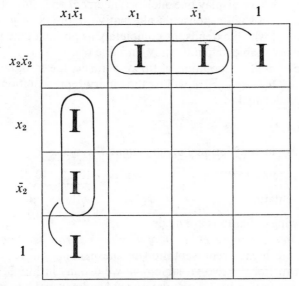

FIG. 3.29. Graphical representation of fuzzy consensus result on a two-variable map $\alpha x_i \bar{x}_i + \bar{\alpha} x_i \bar{x}_i = x_i \bar{x}_i$.

Note: $\text{CF}(x_2 \bar{x}_2 x_1) + \text{CF}(x_2 \bar{x}_2 \bar{x}_1) = \text{CF}(x_2 \bar{x}_2)$
          0.2   +   0.2   =   0.4

Another interesting property of the fuzzy map follows from the application of the fuzzy consensus result

$$\alpha x_i \bar{x}_i \beta + \bar{\alpha} x_i \bar{x}_i \beta = x_i \bar{x}_i \beta. \tag{3.9.1}$$

In terms of the map this result means that we can replace the two squares representing $x_i \bar{x}_i (x_j + \bar{x}_j)$ with the square $x_i \bar{x}_i$ as shown in Fig. 3.29. We note that in this operation the (total) cover fraction is unchanged.

The graphical procedure for minimizing FSF's on the two-variable map involves maximizing the CF of remaining implicants after deletion of lower-CF included implicants. In some cases use of the consensus result is helpful. Figure 3.30 illustrates minimization of some representative functions of two-variables. Before generalizing to more variables we note that the lowermost row of the two-variable map forms a one-variable fuzzy map. This concept will be useful to us in considering maps of N variables where N is odd.

When considering functions of N variables, the basic two-variable map must be enlarged to 4^N squares to provide for all meaningful implicants of the N variables. There are several plausible ways to arrange such a map. In § 3.5 we display one such arrangement and show its use in minimization. We shall espouse an alternative, more clearly structured arrangement which extends, as completely as possible, the properties of the two-variable map into an N variable one.

In order to simplify denotation of fuzzy phrases, the *hexadecimal representation of a fuzzy phrase* will be the binary representation converted into hexadecimal.

Example 3.9.1:

Fuzzy phrase in conventional arrangement: $x_1 x_2 \bar{x}_2 x_3 \bar{x}_4$

Binary representation: $\underbrace{10\ 11}\quad\underbrace{10\ 01}$

Hex representation: B   9

Also the following definitions are made.

(1) The *quaternary representation of a column or row heading* on a fuzzy map is the binary representation of the heading converted to quaternary form. For the two-variable map we display in Fig. 3.31 an equivalent map notation including quaternary headings and hex representations of the fuzzy phrases associated with each square.

(2) The *dominant square* of a fuzzy map is the square representing the phrase 1. This phrase has the hex representation 0.

To construct a map of a function of N variables (where N is even) we let the first (lowest-subscript) $N/2$ variables form column headings. The remaining variables form row headings. For odd N, we shall conventionally have one more column variable than row variable. Further, in construction of the N-variable map we shall treat whole one- or two-variable maps as though they were squares of a map of $N-1$ or $N-2$ variables, respectively. These subsets of the logical space will be called *submaps*. Maps for $N=3$ and $N=4$ are shown in Figure

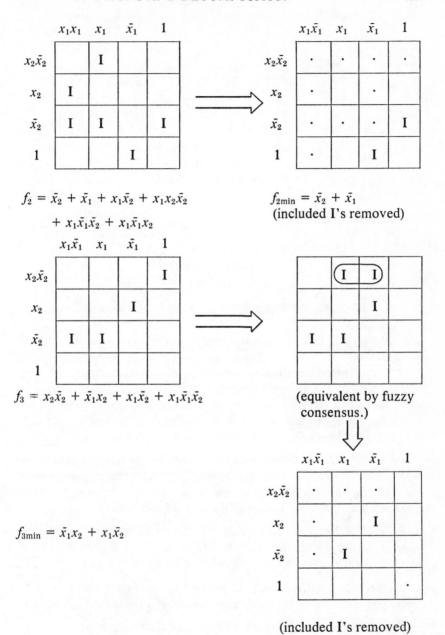

$f_2 = \bar{x}_2 + \bar{x}_1 + x_1\bar{x}_2 + x_1x_2\bar{x}_2$
$\quad + x_1\bar{x}_1\bar{x}_2 + x_1\bar{x}_1x_2$

$f_{2min} = \bar{x}_2 + \bar{x}_1$
(included I's removed)

$f_3 = x_2\bar{x}_2 + \bar{x}_1x_2 + x_1\bar{x}_2 + x_1\bar{x}_1\bar{x}_2$

(equivalent by fuzzy consensus.)

$f_{3min} = \bar{x}_1x_2 + x_1\bar{x}_2$

(included I's removed)

FIG. 3.30. Minimization on a two-variable fuzzy map.

	3	2	1	0	x_1
3	F	B	7	3	
2	E	A	6	2	
1	D	9	5	1	
0	C	8	4	0	

x_2

Example: hex   6
           binary   0110
           phrase $\bar{x}_1\, x_2$
           column heading   1
           row heading   2

FIG. 3.31. A compact notation for the two-variable map.

3.32. N-variable fuzzy maps constructed in this manner will be called S-maps.

We now introduce some important properties of S-maps. We first note that, since the hex representation of phrases on S-maps proceeds by columns from lower right to upper left, the dominant square of the entire S-map is designated hex 0, while the dominant square of any submap is the lowest-numbered square in that submap.

Lemma 3.9.1. A submap on an S-map subsumes (is included by) a second submap if the phrase represented by the dominant square of the first submap subsumes (includes all the literals of) the phrase represented by the dominant square of the second submap.

Proof. Each square of a submap is included by the dominant square (the CF of the dominant square is 1.0). Thus if the dominant square subsumes some other square, the entire submap is included.   Q.E.D.

Theorem 3.9.1. Every subset (pattern of implicant squares) of a submap is included by the dominant square of a subsumed submap.

Proof. A pattern of implicant squares of a submap indicates a disjunction of several fuzzy phrases which is included by the submap's dominant square. If a fuzzy phrase α represents the dominant square of

	x_1=3				x_1=2				x_1=1				x_1=0			
x_2	3	2	1	0	3	2	1	0	3	2	1	0	3	2	1	0
3	3F	3B	37	33	2F	2B	27	23	1F	1B	17	13	0F	0B	07	03
2	3E	3A	36	32	2E	2A	26	22	1E	1A	16	12	0E	0A	06	02
1	3D	39	35	31	2D	29	25	21	1D	19	15	11	0D	09	05	01
0	3C	38	34	30	2C	28	24	20	1C	18	14	10	0C	08	04	00

x_3

x_3	x_4	x_1=3				x_1=2				x_1=1				x_1=0			
	x_2	3	2	1	0	3	2	1	0	3	2	1	0	3	2	1	0
3	3	FF	EF	DF	CF	BF	AF	9F	8F	7F	6F	5F	4F	3F	2F	1F	0F
	2	FE	EE	DE	CE	BE	AE	9E	8E	7E	6E	5E	4E	3E	2E	1E	0E
	1	FD	ED	DD	CD	BD	AD	9D	6D	7D	6D	5D	4D	3D	2D	1D	0D
	0	FC	EC	DC	CC	BC	AC	9C	8C	7C	6C	5C	4C	3C	2C	1C	0C
2	3	FB	EB	DB	CB	BB	AB	9B	8B	7B	6B	5B	4B	3B	2B	1B	0B
	2	FA	EA	DA	CA	AA	AA	9A	8A	7A	6A	5A	4A	3A	2A	1A	0A
	1	F9	E9	D9	C9	B9	A9	99	89	79	69	59	49	39	29	19	09
	0	F8	E8	D8	C8	B8	A8	98	88	78	68	58	48	38	28	18	08
1	3	F7	E7	D7	C7	B7	A7	97	87	77	67	57	47	37	27	17	07
	2	F6	E6	D6	C6	B6	A6	96	86	76	66	56	46	36	26	16	06
	1	F5	E5	D5	C5	B5	A5	95	85	75	65	55	45	35	25	15	05
	0	F4	E4	D4	C4	B4	A4	94	84	74	64	54	44	34	24	14	04
0	3	F3	E3	D3	C3	B3	A3	93	83	73	63	53	43	33	23	13	03
	2	F2	E2	D2	C2	B2	A2	92	82	72	62	52	42	32	22	12	02
	1	F1	E1	D1	C1	B1	A1	91	81	71	61	51	41	31	21	11	01
	0	F0	E0	D0	C0	B0	A0	90	80	70	60	50	40	30	20	10	00

$x_3 x_4$

FIG. 3.32. S-maps for three and four variables.

the submap including the pattern $(\rho_1 + \rho_2 + \ldots + \rho_n)\alpha$ and α subsumes another phrase β, where β represents the dominant square of another submap, then

$$(\rho_1 + \rho_1 + \ldots + \rho_n)\alpha \subseteq \alpha \subseteq \beta \rightarrow (\rho_1 + \rho_2 + \ldots + \rho_n)\alpha \subseteq \beta.$$
(3.9.2)

Theorem 3.9.2. Any subset (pattern of implicant squares) of a submap is included by the same pattern on a subsumed submap.

Proof. Let the first pattern again be represented as a disjunction of phrases

$$(\rho_1 + \rho_2 + \ldots + \rho_n)\alpha$$

where α represents the dominant square of the first submap. Let β represent the dominant square of a second submap. If $\alpha \subseteq \beta$ then

$$(\rho_1 + \rho_2 + \ldots + \rho_n)\alpha \subseteq (\rho_1 + \rho_2 + \ldots + \rho_n)\beta,$$
(3.9.3)
Q.E.D.

These properties of S-maps mean that one may use the same rules for manipulation on 2-variable maps and N-variable maps. On S-maps entire submap-sized patterns behave as single implicants do on the two-variable map. Further, we may extend the concept of a submap to treat larger subsets of the logical space (say $n \times m$ where $N > nm$) as submaps, allowing us to manipulate large patterns of implicants by our usual rules.

To illustrate the use of these concepts in manipulation of FSF's, we shall first demonstrate the use of an S-map to minimize an FSF of four variables. In terms of the map minimization requirements mean that we must end with the fewest possible implicant symbols (**I**) on the map and maximizing the cover fraction of each **I**, since

$$\text{CF} = \frac{2^{n_0}}{4^N}$$

where N is the number of variables on the map and n_0 is the number of 0's in binary representation (each 0 represents a "missing" literal). The procedure uses the rules of inclusion and consensus as developed above.

The example of graphical minimization is exhibited in Fig. 3.33. We begin with the FSF of four variables represented in hexadecimal as

$$f_4(x_1,x_2,x_3,x_4) = \quad 1B + 22 + 2B + 54 + 5C + 64 + 6C + 85$$

$$+ 8B + 91 + AB + CC + D6 + D9 + DD + DF$$

$$+ EB + EE.$$

(a)

(b)

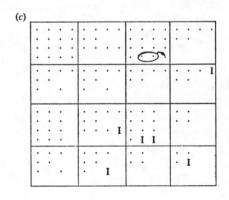

(c)

FIG. 3.33. Minimization on a four-variable S-map. (For clarity only submap boundaries are shown.) (a) Notes: (1) the dominant square CC of the upper-left corner submap includes the whole submap; (2) since 5C and 6C are in a 3 row, consensus means 5C + 6C = 4C; (3) we may similarly replace 2B + 1B = 0B. (b) Included implicants within submaps removed.
Notes: (4) 4C includes all of CC's submap, but since 4C = 5C + 6C, 4C is included in 64 + 54; (5) 0B includes 8B and EB, i.e. EB ⊆ 8B ⊆ 0B; (6) D2 ⊆ 54; (7) D9 ⊆ 91. (c) Minimized function shown as I's and included implicant squares as ·'s:

$$f_{4\,\text{min}} = 0B + 22 + 54 + 64 + 85 + 91.$$

Operations are performed as indicated in Fig. 3.33b to eliminate most of the phrases, and end with the equivalent, minimized function

$$f_{4\min}(x_1,x_2,x_3,x_4) = 0B + 22 + 54 + 64 + 85 + 91$$

as shown in Fig. 3.33c.

An important problem in the realm of FSF's is that of decomposition. That we may use S-maps to perform FSD decomposition and gain insight into this process is our topic here. This view of decomposition will differ significantly from that of §3.8 in which the operators $\oplus$ and $\odot$ are the only ones considered for use in the bound-variable function $\psi(A)$. This approach may be briefly stated in this way. We inspect maps of the function in disjunctive normal form for the existence of patterns of implicant squares related in the same fashion that two-variable functions are related under DeMorgan's laws. If such patterns are found, we begin construction of A and $\psi(A)$ using the indicated variables and operators. We may iterate the procedure to extend A and increase the complexity of $\psi(A)$.

We begin by choosing two of the N variables as candidates for membership in A, the set of bound variables. We then map $F(x_1,x_2,\ldots,x_n)$ (in d.n.f.) on an S-map of N-variables, of which our two candidates appear as the outer row and column variables. If by valid manipulation of the map as described above, we can make sub-map patterns appear as any of the overall forms shown in Fig. 3.34b, the two candidate variables may be admitted to A as bound variables related under the corresponding operations from $\{+, \cdot, \oplus, \odot\}$. If this cannot be done no decomposition is possible under these conditions of bound variables, and another pair of bound-variable candidates must be tried until all pairs have been evaluated. An example of this first step is shown in Fig. 3.35 in which the variables x_1 and x_3 and the operation $\cdot$ (Min) form a function $\psi(x_1,x_3)$. We note that the addition of the phrase $x_2 \bar{x}_2 x_3$ to the function F would have prevented this decomposition.

After finding a workable $\psi(A)$ we may iterate the procedure by defining a new variable $x_\psi = \psi(A)$ and mapping an equivalent form of F which is actually a decomposition $f\{\psi(A),B\}$ on an $N-1$ variable S-map. If we can again find a decomposition in which x_ψ is a bound variable, we extend A and define a new $\psi(A)$ with three bound variables. The iteration may be repeated until all combinations of variables and operations have been tried. An example of such a decomposition process is exhibited in Fig. 3.36. This decomposition procedure

$$\alpha(x_i, x_j) = x_i x_j$$
$$\bar{\alpha} = \bar{x}_i + \bar{x}_j$$

$$\beta(x_i, x_j) = \bar{x}_i \bar{x}_j$$
$$\beta = x_i + x_j$$

$$(a)$$

$$\gamma(x_i, x_j) = \bar{x}_i \bar{x}_j$$
$$\bar{\gamma} = x_i + \bar{x}_j$$

$$\delta(x_i, x_j) = x_i \bar{x}_j$$
$$\delta = x_i + x_j$$

$$\epsilon(x_i, x_j) = x_i \oplus x_j = x_i \odot \bar{x}_j = \bar{x}_i \odot x_j$$
$$\bar{\epsilon} = x_i \odot x_j = x_i \oplus \bar{x}_j = \bar{x}_i \oplus x_j$$

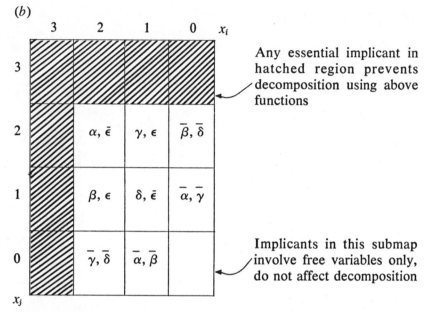

FIG. 3.34. Graphical representation of submap-sized regions occupied by implicant patterns in FSD decompositions. The functions α, β, γ, δ, and ϵ are possible decomposition functions in an FSD decomposition form

$$f\big[\psi(A), B\big] \text{ where } A = \{x_i, x_j\}.$$

can produce all FSD decompositions which do not contain $x_i \bar{x}_i$ in the bound-variable set. Fuzzy complex disjunctive (FCD) decompositions may also be recognized when two or more eligible but disjoint bound-variable sets are found.

It should be emphasized, however, that the task of constructing a complete decomposition theory of fuzzy structures is one of very considerable magnitude, and that what we had to say about decomposition of fuzzy functions is merely a first step toward devising a conceptual framework for dealing with decompositions of such structures in future work.

Exercises

(1) With the aid of a fuzzy map derive minimal disjunctive expressions for each of the following fuzzy switching functions:

(a) $f_1(x,y) = \sum (0, 1, 2, 7, 9, 14)$

(b) $f_2(x,y,z) = \sum (6, 7, 14, 15, 22, 24, 27, 28)$

(c) $f_3(w,x,y,z) = \sum (31, 32, 33, 34, 79, 80, 91, 99)$.

(2) Determine the minimal disjunctive expression for

$$f(w,x,y,z) = \sum (34, 127, 190, 230, 233)$$

$$+ \sum_\Phi (128, 154, 234, 239).$$

(3) Use the consensus method to simplify the functions in problems (1) and (2).

(4) For the function $T(w,x,y,z) = \sum (6, 7, 41, 45, 46, 53, 58, 60)$:

    (a) show the fuzzy map
    (b) find all fuzzy prime implicants and indicate which are essential
    (c) find a minimal expression for T.

(5) The upper bound on the number of fuzzy switching functions of n variables is 2^{4^n}. Clearly the number of *minimized* fuzzy switching functions of n variables is much smaller. Can you find a better bound for this set? *Hint:* For $n = 2$, $2^{4^n} = 65,536$ whereas there are only 84 minimized fuzzy functions of two variables.

(6) Let

$$f_1(x,y,z) = (\bar{x} + y)(x + \bar{y} + \bar{z})$$

and

$$f_2(x,y,z) = (x + y)(\bar{x} + \bar{y}).$$

Discuss the limiting value of F where $F = f_1 + f_2$. What interesting deductions can you make regarding the fuzzy function F?

(7) Repeat problem (6) for

$$g_1(x,y,z) = yz + \bar{y}\bar{z}$$
$$g_2(x,y,z) = \bar{x}\bar{y}z + y\bar{z}$$

and

$$G = g_1 g_2.$$

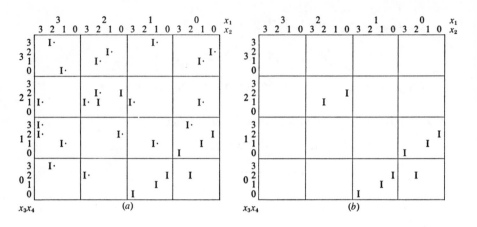

FIG. 3.35. FSD decomposition on a four-variable S-map. Note: Adding 38 to F_5 will prevent decomposition.

(a) The original function:

$$F_5(x_1,x_2,x_3,x_4) = 06 + 0E + 15 + 19 + 1D + 22$$
$$+ 27 + 34 + 42 + 51 + 55 + 5F$$
$$+ 70 + 79 + 86 + 8A + 9E + A9$$
$$+ AA + AD + B2 + B9 + D5 + DC$$
$$+ E3 + EF + F6 + F7 + F9$$

The following implicants may be removed without changing F:

(1) $9E \subseteq 0E \subseteq 06;\ 27 \subseteq 06;\ 86 \subseteq 06$
(2) $AD \subseteq 1D \subseteq 19 \subseteq 15$
(3) $EF \subseteq 5F \subseteq 55 \subseteq 51;\ D5 \subseteq 55 \subseteq 51$
(4) $79 \subseteq 70$
(5) $DC = DD + DE$ (consensus)
        $DC \subseteq 15 + 06$
(6) $F7 \subseteq F6 \subseteq 22;\ B2 \subseteq 22$
(7) $E3 \subseteq 22$
(8) $F9 \subseteq B9$
(9) $AA \subseteq 8A$
(10) $B9 \subseteq A9.$

(b) An equivalent function reduced to show

$$F = f[\psi(A), B];\ A = \{x_1,x_3\}\quad B = \{x_2,x_4\}$$
$$\psi(A) = \bar{x}_1 + \bar{x}_3$$

$$F_5 = (x_4 + x_2\bar{x}_4)\left[\psi(A)\right] + (x_4 + \bar{x}_2\bar{x}_4 + x_2\bar{x}_2)\left[\overline{\psi(A)}\right] + x_2x_4.$$

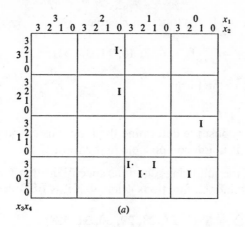

FIG. 3.36. Iterated four-variable decomposition with three bound variables. The function

$$F_6(x_1,x_2,x_3,x_4) = \bar{x}_2\bar{x}_3x_4\bar{x}_4 + x_2x_4 + \bar{x}_1\bar{x}_2x_4\bar{x}_4 + \bar{x}_1x_2x_4$$
$$+ \ \bar{x}_1x_2\bar{x}_2x_4\bar{x}_4 + x_1x_3x_4 + x_1x_3\bar{x}_3x_4$$
$$= 17 + 22 + 53 + 62 + 73 + 8A + 8E$$

(1) Since $62 \subseteq 22$, $73 \subseteq 53$, $8E \subseteq 8A$ remove 62, 73, 8E

(2) Now $\psi(x_1,x_3) = x_1 \cdot x_3$ and
$F_6 = \psi(x_1,x_3)x_4 + \overline{\psi}(x_1,x_2)\bar{x}_2x_4\bar{x}_4 + x_2x_4$

(3) Define $x_\psi = x_1 \cdot x_3$ and remap F_6 on the three-variable S-map:

$F_6(x_\psi,x_2,x_4) = x_\psi x_4 + \bar{x}_\psi\bar{x}_2x_4\bar{x}_4 + x_2x_4.$

(4) Recognize additional decomposition with new
$\psi(A) = \bar{x}_\psi\bar{x}_2$
$\qquad = (\bar{x}_1 + \bar{x}_3)\bar{x}_2$

$\qquad F_6 = \psi(x_\psi,x_2)x_4\bar{x}_4 + \overline{\psi}(x_\psi,x_2)x_4.$

(5) $F_6(x_1,x_2,x_3,x_4) = \psi(x_1,x_2,x_3)x_4\bar{x}_4 + \overline{\psi}(x_1,x_2,x_3)x_4$
where $\psi(x_1,x_2,x_3) = \bar{x}_1\bar{x}_2 + \bar{x}_3\bar{x}_2.$

(8) Can the normal fuzzy function F_n given by

$$F_n(v,w,x,y,z) = \sum (1, 3, 6, 7, 10, 11, 30, 31)$$

be decomposed to the form

$$F\left[\Phi(w,y,z),v,x\right] = ?$$

If your answer is positive determine the functions F and Φ; otherwise explain why F_n is not decomposable in the form above.

(9) For each of the following functions specify the don't care combinations and determine the functions ψ and Φ if it is FSD decomposable:

(a) $f(w,x,y,z) = \sum (2, 27, 28, 29, 30, 31, 128)$

$$+ \sum_\phi (34, 229, 230, 234)$$

(b) $g(w,x,y,z) = \sum (154, 190, 218, 228, 233)$

$$+ \sum_\phi (31, 33, 35, 37, 38, 116, 234).$$

(10) Compare the decomposition of N-variable fuzzy switching functions by the two different techniques discussed in this chapter.

(11) Prove that if a negated fuzzy switching function $\bar{f}$ is a single phrase, the complement of each literal is a fuzzy prime implicant of f and these are the only FPI's of f.

(12) The consensus operation is not easy to perform, unless one is using a graphical (N-tabular, pictorial, etc.) technique like the K-map. Devise another means with which minimization of fuzzy switching can be performed, utilizing the consensus concept.

CHAPTER 4

Fuzzy Automata and Languages

4.1. INTRODUCTION

The theory of fuzzy sets discussed in Chapter 2 has been applied to systems theory[1,2] with particular emphasis on automata[3,4,5] and languages[6,7]. In this chapter fuzzy matrices are first introduced based on which fuzzy automata and languages are defined. We then demonstrate the use of regular expressions in describing finite fuzzy automata and regular fuzzy languages, and introduce a normal form for productions of a regular fuzzy grammar when the "max(min)" rule is used. Finally, we explore the relationships between formal languages and λ-fuzzy languages.

A brief summary of finite automata and language theory and its extension to finite fuzzy automata and languages and an introduction to its extension to finite fuzzy automata and languages are given below.

A finite automaton FA is a system analytically describable by the five-tuple FA $= (\Sigma, Q, \delta, S_0, T)$, where Σ is the finite input alphabet, Q is the finite state set, $\delta: Q \times \Sigma \to Q$ is the state transition map, $S_0(\in Q)$ is the initial state, and $T(\subset Q)$ is the set of final (accepting) states. The domain of δ is extended to $Q \times \Sigma^*$, where Σ^* is the free monoid over Σ. In operation the FA is started in state S_0 with input string $x \in \Sigma^*$. If the FA halts in a state in T after reading x, then x is accepted;

171

otherwise x is rejected. Note that we can define an output map $\omega: Q \rightarrow \{0,1\}$ such that

$$\omega(F) = \begin{cases} 1 \text{ if } F \in T \\ 0 \text{ otherwise;} \end{cases}$$

that is ω is the characteristic function of the set T. Then x is accepted iff (if and only if) after reading x the FA halts in a state with output 1. The set of strings in Σ^* accepted by the FA is denoted by $T(FA)$.

A regular (type-3) grammar, RG, is a system $RG = (V_N, V_T, S_0, P)$ where V_N is the finite set of non-terminals, V_T is the finite set of terminals, $S_0 (\in V_N)$ is the initial symbol, and P is the finite set of productions. It is required that $V_N \cap V_T = \emptyset$, the null set, and that elements of P be of the form $A \rightarrow aB$ or $A \rightarrow a$ where $A, B \in V_N$ and $a \in V_T$. A word in the regular language $L(RG)$ generated by RG is a string in V_T^* derived from S_0 by use of a finite number of productions.

A finite automaton solves the decision problem for a regular language; for a regular grammar RG (finite automaton FA) there is a corresponding finite automaton FA (regular grammar RG) such that $L(RG) = T(FA)$. The relationship between regular languages and regular expressions is also one to one. RL is a regular language iff $RL = L(RG)$ for a regular grammar RG, iff $RL = T(FA)$ for a finite automaton FA, or iff RL is described by a regular expression. The finite automaton model is generalized to a finite fuzzy automaton FFA by replacing δ by $\hat{\delta}: Q \times \Sigma \times [0,1] \rightarrow Q$. The $\hat{\delta}$ map associates with each transition a number in the closed interval $[0,1]$ which is a fuzzy set membership grade. Similarly, the definition of a regular grammar is generalized to a regular fuzzy grammar RFG by associating with each production a fuzzy set membership grade. Productions in an RFG are of the form $A \xrightarrow{\sigma} aB$ or $A \xrightarrow{\sigma} a$ where $A, B \in V_N$, $a \in V_T$, and $0 < \sigma \leq 1$. In an RFG (or FFA) those productions (or transitions) not specified are understood to have membership grade 0.

Just as a language over alphabet Σ is defined to be a set in Σ^*, so a fuzzy language FL over Σ is defined to be a fuzzy set in Σ^* [8]. A string x of Σ^* is not necessarily either "in the fuzzy language FL" or "not in the fuzzy language FL"; rather x has a membership grade $\mu_{FL}(x)$, $0 \leq \mu_{FL}(x) \leq 1$, which measures its degree of membership in FL.

There is a one-to-one correspondence between an FFA as an acceptor and an RFG as a generator of a regular fuzzy language RFL. The fuzzy grammar is said to be unambiguous if, for a string x with non-zero membership grade in RFL, there is only one derivation of x. The usual determination of the membership grade $\mu_{RFL}(x)$ for such a derivation is to take the minimum ("min") of the membership grades of all productions used. If there is more than one derivation of x, that is if the grammar is ambiguous, then the actual membership grade $\mu_{RFL}(x)$ is taken to be the maximum ("max") of the membership grades obtained from all derivations of x. This is the "max(min)" rule.

Example 4.1.1. Consider the derivations $S_0 \xrightarrow{0.5} aA \xrightarrow{0.5} ab$ and $S_0 \xrightarrow{0.7} aB \xrightarrow{0.4} ab$. The membership grade of string ab is

$$\mu_{RFL}(ab) = \max \left[\min(0.5, 0.5), \min(0.7, 0.4) \right]$$

$$= \max \left[0.5, 0.4 \right]$$

$$= 0.5.$$

The context-sensitive fuzzy languages are recursive; that is, there is an algorithm for computing the membership grade of a string in a context-sensitive fuzzy language when the "max(min)" derivation rule is used. Since the regular fuzzy languages are in particular context sensitive, regular fuzzy languages are recursive. Furthermore, the set M of possible non-zero membership grades in an RFL is finite because (i) the set P of productions in an RFG is finite and (ii) the max and min operations cannot introduce a membership grade not already assigned to some production. These facts are used to relate regular expressions to RFL's.

4.2. FUZZY MATRICES

Similar to a stochastic process, we define a fuzzy process as follows.

Definition 4.2.1 A (finite) process in discrete time with a discrete state space $Q = \{q_1, q_2, \ldots, q_n\}$ is called a (finite) *fuzzy* process if it satisfies the following conditions.

(1) The matrix $\mathbf{F}$ which describes the state transition has the following form

$$\mathbf{F} = \begin{array}{c} \\ q_1 \\ q_2 \\ \vdots \\ q_r \end{array} \begin{array}{cccc} q_1 & q_2 & \cdots & q_r \\ \left[\begin{array}{cccc} f_{11} & f_{12} & \cdots & f_{1r} \\ f_{21} & f_{22} & \cdots & f_{2r} \\ \vdots & \vdots & & \vdots \\ f_{r1} & f_{r2} & \cdots & f_{rr} \end{array} \right] \end{array}$$

where $0 \leqslant f_{ij} \leqslant 1$ denotes the grade of membership of state transition from state q_i to state q_j. This matrix will be called the *fuzzy state transition matrix* of the fuzzy process.

(2) Let $\mathscr{A}$ be a fuzzy set defined on S, and $w^{(0)}_{\mathscr{A}} = \left[\eta_{q_1}^{'(0)'}\eta_{q_2}^{(0)'} \ldots \eta_{q_r}^{(0)}\right]$ be a row vector, called the *initial state designator of $\mathscr{A}$*, where η is the grade of membership of s_i with respect to $\mathscr{A}$. Then the state designator of $\mathscr{A}$ at $t = n$, $w^{(n)}_{\mathscr{A}} = \left[\eta_{q_1}^{(n)}\eta_{q_2}^{(n)} \ldots \eta_{q_r}^{(n)}\right]$ is obtained by

$$w^{(n)}_{\mathscr{A}} = w^{(0)}_{\mathscr{A}} \circ \underbrace{\{\mathbf{F} \circ \mathbf{F} \circ \ldots \circ \mathbf{F}\}}_{(n-1) \text{ operations}} = w^{(0)}_{\mathscr{A}} \circ \mathbf{F}^n = w^{(0)}_{\mathscr{A}}\left[f_{ij}^{(n)}\right]$$

where

$$f_{ij}^{(n)} = \max_{\substack{\text{over all parallel} \\ \text{paths from } q_i \text{ to } q_j}} \quad \min_{\substack{\text{over all series} \\ \text{paths from } q_i \text{ to } q_j}} \quad \{\mu_{ij}\mid$$

μ_{ij} = the set of grades of membership of state transitions of a path from q_i to $q_j.\}$

Remarks

(1) The fuzzy state transition matrix $\mathbf{F}$ of a fuzzy process resembles somewhat a stochastic matrix of a Markov chain for $0 \leqslant f_{ij} \leqslant 1$; however, in general $\sum_{k=1} f_{ik} \neq 1$, for all i.

(2) The reason we use the max.min.(.) rule in defining $f_{ij}^{(n)}$ is that the state transition from q_i to q_j in a fuzzy process may be considered as water (gas, electricity, traffic flow, etc.) through a water supply system of which the water pipes are series–parallel interconnected.

(3) When f_{ij} takes only two values 0 and 1, the process becomes a non-deterministic process. In addition, only any one element of each row of matrix $\mathbf{F}$ is equal to 1, and the rest of the elements of each row are equal to 0. Then the process becomes a deterministic process.

Definition 4.2.2. A fuzzy process is *stationary* if its state transition matrix is independent of time; otherwise it is *non-stationary*.

In this and the next chapter, a fuzzy process will be referred to as a stationary fuzzy process. We denote by a_{ij} the (i,j)th entry of a fuzzy matrix $\mathbf{A}$, where $0 \leqslant a_{ij} \leqslant 1$.

Example 4.2.1. Let $\mathscr{P}$ be a fuzzy process whose state transition matrix $\mathbf{F}$ is

$$\mathbf{F} = \begin{array}{c} \\ q_1 \\ q_2 \\ q_3 \end{array} \begin{array}{ccc} q_1 & q_2 & q_3 \\ \begin{bmatrix} 0.7 & 0.4 & 0.8 \\ 0.6 & 0.5 & 0.2 \\ 0.3 & 0.9 & 0.1 \end{bmatrix} \end{array}.$$

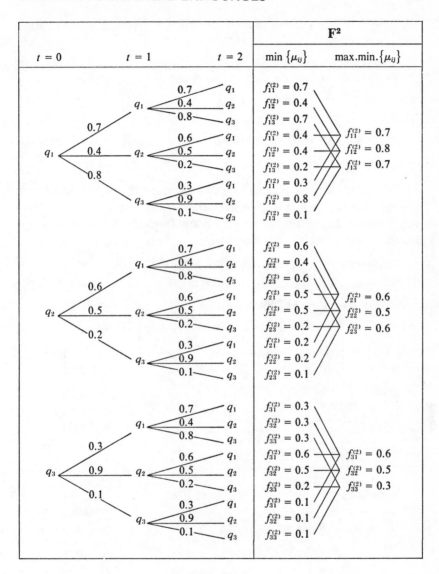

TABLE 4.1

Let $\mathcal{A}$ be a fuzzy set defined on the state space $Q = \{q_1, q_2, q_3\}$, with $w_{\mathcal{A}}^{(0)} = (0.5, 0.7, 0.3)$. The $w_{\mathcal{A}}^{(2)}$ may be found as follows. First of all we want to find $\mathbf{F^2}$. The method of obtaining $\mathbf{F^2}$ is shown in Table 4.1. We obtain

$$\mathbf{F^2} = \begin{bmatrix} 0.7 & 0.8 & 0.7 \\ 0.6 & 0.5 & 0.6 \\ 0.6 & 0.5 & 0.3 \end{bmatrix}.$$

Thus

$$w_{\sim l}^{(2)} = w_{\sim l}^{(0)} \circ \mathbf{F^2} = (0.42, 0.4, 0.42).$$

This may be seen from Figure 4.1.

Just as the stochastic matrix plays the central role in a stochastic process, the fuzzy state transition matrix, or simply fuzzy matrix, plays the central role in a fuzzy process. Some fundamental properties of a fuzzy matrix are given in the following.

Definition 4.2.3. Let A,B be fuzzy matrices. We define $A < B$ iff $a_{ij} \leqslant b_{ij}$.

Definition 4.2.4. Define

$$0 = \begin{bmatrix} 0 & \cdots & 0 \\ \vdots & & \vdots \\ 0 & \cdots & 0 \end{bmatrix} \qquad E = \begin{bmatrix} 1 & \cdots & 1 \\ \vdots & & \vdots \\ 1 & \cdots & 1 \end{bmatrix}$$

Definition 4.2.5.

$$C = A \circ B \Leftrightarrow c_{ij} = \max_k \min(a_{ik}, b_{kj})$$

$$I = \begin{bmatrix} m_{ij} \end{bmatrix} \text{ where } m_{ij} = \begin{cases} 1 \text{ if } i = j \\ 0 \text{ if } i \neq j \end{cases}$$

$$A^{m+1} = A^m \circ A, \quad A^0 = I$$
$$C = A * B \Leftrightarrow c_{ij} = \min_k \max(a_{ik}, b_{kj})$$

$$I' = \begin{bmatrix} m'_{ij} \end{bmatrix} \text{ where } m'_{ij} = \begin{cases} 0 \text{ if } i = j \\ 1 \text{ if } i \neq j \end{cases}$$

$$B^{m+1} = B^m * B, \quad B^0 = I'$$

Property 1. Let A be a fuzzy matrix. Then $0 < A < E$.

Property 2. Define $\mathbf{A}^m = \underbrace{\mathbf{A} \circ \mathbf{A} \circ \ldots \circ \mathbf{A}}_{m \mathbf{A}\text{'s}}$. If $\mathbf{A}$ is a fuzzy matrix, so is $\mathbf{A}^m$.

The proof of this property is evident and thus may be omitted.

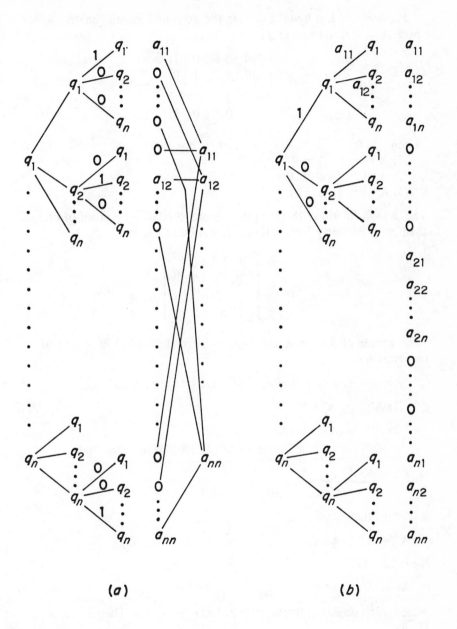

(a)   (b)

FIG. 4.1.

Property 3. Let **0** and **I** denote the zero and identity matrices, respectively. Let the matrix **J** be

$$\mathbf{J} = \begin{bmatrix} 0 & 0 \ldots 0 & 1 \\ 0 & 0 \ldots 1 & 0 \\ \vdots & \vdots & \vdots & \vdots \\ 1 & 0 \ldots 0 & 0 \end{bmatrix}.$$

Then

(1) $\mathbf{A} \circ \mathbf{0} = \mathbf{0} \circ \mathbf{A} = \mathbf{0}$

(2) $\mathbf{A} \circ \mathbf{I} = \mathbf{I} \circ \mathbf{A} = \mathbf{A}$

(3) $\mathbf{A} \circ \mathbf{J} = \mathbf{J} \circ \mathbf{A} = \mathbf{A}.$

The proofs for these three equalities are similar. The proof of one of them, say Property 3(2), suffices. Let **A** be as shown.

$$\mathbf{A} = \begin{bmatrix} a_{11} & a_{12} \ldots a_{1n} \\ a_{21} & a_{22} \ldots a_{2n} \\ \vdots & \vdots & \vdots \\ a_{n1} & a_{n2} \ldots a_{nn} \end{bmatrix}.$$

The proofs of $\mathbf{A} \circ \mathbf{I} = \mathbf{A}$ and $\mathbf{I} \circ \mathbf{A} = \mathbf{A}$ are shown in Fig. 4.1*a* and *b*, respectively.

Property 4. Let m and n be positive integers. Then

(1) $\mathbf{A}^{m+n} = \mathbf{A}^m \circ \mathbf{A}^n$

(2) $\mathbf{A}^{mn} = (\mathbf{A}^m)^n = (\mathbf{A}^n)^m$

Proof: From the definition of $\mathbf{A}^p$, where p is an integer,

$$\mathbf{A}^p = \underbrace{\mathbf{A} \circ \mathbf{A} \circ \ldots \circ \mathbf{A}}_{p\,\mathbf{A}\text{'s}}$$

it follows that

$$\mathbf{A}^p = \mathbf{A}^{p-1} \circ \mathbf{A}.$$

By induction,

$$\mathbf{A}^p = \mathbf{A}^{p-n} \circ \mathbf{A}^n$$

where n is a positive integer, $p > n$. Let $p = m + n$. Then $p - n = m$. Hence

$$\mathbf{A}^{m+n} = \mathbf{A}^m \circ \mathbf{A}^n.$$

Suppose p is equal to a product of two integers m and n, i.e. $p = mn$. Then $\mathbf{A}^p$ may be expressed

$$\mathbf{A}^p = \underbrace{\underbrace{\mathbf{A} \circ \mathbf{A} \circ \ldots \circ \mathbf{A}}_{m\mathbf{A}\text{'s}} \cdots \underbrace{\mathbf{A} \circ \mathbf{A} \circ \ldots \circ \mathbf{A}}_{m\mathbf{A}\text{'s}}}_{n \text{ such terms}} = (\mathbf{A}^m)^n$$

This completes the proof.

Property 5. Let $\mathbf{A}$, $\mathbf{B}$, and $\mathbf{C}$ be any fuzzy matrices. Then

$$\mathbf{A} \circ (\mathbf{B} \circ \mathbf{C}) = (\mathbf{A} \circ \mathbf{B}) \circ \mathbf{C}.$$

Proof. Let us consider a typical element in $\mathbf{L} = \mathbf{A} \circ (\mathbf{B} \circ \mathbf{C}) = (l_{ij})$. Let $\mathbf{D} = \mathbf{B} \circ \mathbf{C} = (d_{ij})$. Then

$$l_{ij} = \max_k \left(\min(a_{i,k}, d_{k,j}) \right)$$

$$= \max \left(\min_k \left[a_{i,k}, \max(\min_v [b_{kv}, c_{vj}]) \right] \right).$$

Similarly, if we define

$$\mathbf{E} = (\mathbf{A} \circ \mathbf{B}) = (e_{ij})$$

then the corresponding element at the ith row and j^{th} column of

$$\mathbf{R} = (\mathbf{A} \circ \mathbf{B}) \circ \mathbf{C} = (r_{ij})$$

will be

$$r_{ij} = \max \left(\min_k \left[e_{ik}, c_{kj} \right] \right)$$

$$= \max \left(\min_k \left[\max \left(\min_v [a_{iv}, b_{vk}] \right), c_{kj} \right] \right).$$

Property 6. Let $\mathbf{A}$, $\mathbf{B}$, $\mathbf{C}$, and $\mathbf{D}$ be fuzzy matrices. Let $\mathbf{A} = [a_{ij}]$ and $\mathbf{B} = [b_{ij}]$. Then, if $\mathbf{A} < \mathbf{B}$ and $\mathbf{C} < \mathbf{D}$, $\mathbf{A} \circ \mathbf{C} < \mathbf{B} \circ \mathbf{D}$.

Proof. The proof is somewhat similar to that of Property 4. Consider the ith row and j^{th} column element of $\mathbf{A} \circ \mathbf{C}$ and $\mathbf{B} \circ \mathbf{D}$, denoted by e_{ij} and f_{ij}. Then

$$e_{ij} = \max \left(\min_k \left[a_{ik}, c_{kj} \right] \right)$$

$$f_{ij} = \max \left(\min_k \left[b_{ik}, d_{kj} \right] \right).$$

Since $a_{ik} \leq b_{ik}$ and $c_{kj} < d_{kj}$ for all k,

$$\min_k \left[a_{ik}, c_{kj} \right] \leq \min_k \left[b_{ik}, d_{kj} \right] \text{ for all } k.$$

Hence $e_{ij} \leq f_{ij}$. This completes the proof.

Example 4.2.2. Let

$$A = \begin{bmatrix} 0.1 & 0.2 \\ 0.4 & 0.3 \end{bmatrix} \qquad B = \begin{bmatrix} 0.2 & 0.3 \\ 0.5 & 0.4 \end{bmatrix}$$

$$C = \begin{bmatrix} 0.5 & 0.4 \\ 0.3 & 0.6 \end{bmatrix} \qquad D = \begin{bmatrix} 0.7 & 0.9 \\ 0.6 & 0.8 \end{bmatrix}$$

Where **A** <**B** and **C**<**D**

A ∘ C

B ∘ D

$$A \circ C = \begin{bmatrix} 0.2 & 0.2 \\ 0.4 & 0.4 \end{bmatrix} \qquad\qquad B \circ D = \begin{bmatrix} 0.3 & 0.3 \\ 0.5 & 0.4 \end{bmatrix}$$

Hence **A ∘ C** < **B ∘ D.**

All the above six fundamental properties of fuzzy matrices also hold for the operation $*$.

Theorem 4.2.1.

(1) $0 < A < E$

(1') $0 < B < E$

(2) $A \circ A \circ \ldots \circ A$ is a
    fuzzy matrix

(2') $A * A * \ldots * A$ is a
    fuzzy matrix

(3a) $A \circ 0 = 0 \circ A = 0$

(3'a) $B * E = E * B = E$

(3b) $A \circ I = I \circ A = A$

(3'b) $B * I' = I' * B = B$

(3c) $A \circ J = J \circ A = A$

(3'c) $B * J' = J' * B = B$

(4a) $A^{m+n} = A^m \circ A^n$

(4'a) $B^{m+n} = B^m * B^n$

(4b) $A^{mn} = (A^m)^n = (A^n)^m$

(4'b) $B^{mn} = (B^m)^n = (B^n)^m$

(5) $A \circ (B \circ C) = (A \circ B) \circ C$

(5') $A * (B * C) = (A * B) * C$

(6) if $A < B$ and $C < D$,
    then $A \circ C < B \circ D$

(6') if $A < B$ and $C < D$,
    then $A * C < B * D$

The proofs of the properties $(1')$–$(6')$ are similar to those of the properties (1)–(6).

From Theorem 4.2.1 we have the following.

Theorem 4.2.2. For any $n \times n$ fuzzy transition matrix $F(\sigma)$, the sequence $F(\sigma)$, $F(\sigma^2)$, $F(\sigma^3)$, \ldots is ultimately periodic.

Proof. Let $T = \{f_1, f_2, \ldots, f_l\}$ be the set of all the elements which occur in the matrix $F(\sigma)$; then the number of different matrices which can be obtained by multiplying $F(\sigma)$ is at most l^{n^2} that is finite.

Theorem 4.2.3. If $I < F(\sigma)$, then

$$I < F(\sigma) < F(\sigma^2) < \ldots < F(\sigma^{n-1}) = F(\sigma^n) = F(\sigma^{n+1}) = \ldots$$

Proof. The proof of this theorem is left to the reader as an exercise.

Lemma 4.2.1. (Zadeh). DeMorgan's laws and the distributive laws hold for fuzzy sets, that is, if $\mathcal{A}$, $\mathcal{B}$, $\mathcal{C}$ are fuzzy sets.

$$(\mathcal{A} \cup \mathcal{B})' = \mathcal{A}' \cap \mathcal{B}'$$

DeMorgan's laws

$$(\mathcal{A} \cap \mathcal{B})' = \mathcal{A}' \cup \mathcal{B}'$$

$$\mathcal{C} \cap (\mathcal{A} \cup \mathcal{B}) = (\mathcal{C} \cap \mathcal{A}) \cup (\mathcal{C} \cap \mathcal{B})$$

Distributive laws

$$\mathcal{C} \cup (\mathcal{A} \cap \mathcal{B}) = (\mathcal{C} \cup \mathcal{A}) \cap (\mathcal{C} \cup \mathcal{B})$$

Proof. Let $f_{\mathcal{A}}$, $f_{\mathcal{B}}$, and $f_{\mathcal{C}}$ be the membership functions of fuzzy sets $\mathcal{A}$, $\mathcal{B}$, and $\mathcal{C}$. DeMorgan's laws in terms of fuzzy set operations may be expressed as

$$1 - \max[f_{\mathcal{A}}, f_{\mathcal{B}}] = \min[1 - f_{\mathcal{A}}, 1 - f_{\mathcal{B}}]$$
$$1 - \min[f_{\mathcal{A}}, f_{\mathcal{B}}] = \max[1 - f_{\mathcal{A}}, 1 - f_{\mathcal{B}}]$$

DeMorgan's law

which can be verified to be an identity by testing it for the two possible cases $f_{\mathcal{A}}(x) > f_{\mathcal{B}}(x)$ and $f_{\mathcal{A}}(x) < f_{\mathcal{B}}(x)$.

Similarly, the distributive laws may be expressed as

$$\min[f_{\mathcal{C}}, \max[f_{\mathcal{A}}, f_{\mathcal{B}}]] = \max[\min[f_{\mathcal{C}}, f_{\mathcal{A}}], \min[f_{\mathcal{C}}, f_{\mathcal{B}}]]$$
$$\max[f_{\mathcal{C}}, \min[f_{\mathcal{A}}, f_{\mathcal{B}}]] = \min[\max[f_{\mathcal{C}}, f_{\mathcal{A}}], \max[f_{\mathcal{C}}, f_{\mathcal{B}}]]$$

which can be verified by considering the six cases below.

$$f_{\mathcal{A}}(x) > f_{\mathcal{B}}(x) > f_{\mathcal{C}}(x), \qquad f_{\mathcal{A}}(x) > f_{\mathcal{C}}(x) > f_{\mathcal{B}}(x)$$
$$f_{\mathcal{B}}(x) > f_{\mathcal{A}}(x) > f_{\mathcal{C}}(x), \qquad f_{\mathcal{B}}(x) > f_{\mathcal{C}}(x) > f_{\mathcal{A}}(x)$$
$$f_{\mathcal{C}}(x) > f_{\mathcal{A}}(x) > f_{\mathcal{B}}(x), \qquad f_{\mathcal{C}}(x) > f_{\mathcal{B}}(x) > f_{\mathcal{A}}(x)$$

From the above lemma, we have the following theorem.

Theorem 4.2.4. Let $F = (f_{ij})$ be a fuzzy matrix. Define the complement of a fuzzy matrix as

$$\overline{F} = (1 - f_{ij}).$$

Then

$$(\overline{F \circ F \circ \ldots \circ F}) = \overline{F} * \overline{F} * \ldots * \overline{F}.$$

Proof. First, we show the two-matrix case, i.e.

$$(\overline{F \circ F}) = \overline{F} * \overline{F}$$

Consider an element of $(\overline{F \circ F})$, say at i^{th} row and jth column,

$$1 - \max_{k} \min[f_{ik}, f_{kj}].$$

Using the first DeMorgan law

$$1 - \max_{k} \min[f_{ik}, f_{kj}] = \min[1 - \min(f_{i1}, f_{1j}), \ldots ,$$
$$1 - \min(f_{in}, f_{nj})]$$

By the second DeMorgan law

$$1 - \max_k \min[f_{ik}, f_{kj}] = \min[\max(1 - f_{i1}, 1 - f_{1j}), \ldots,$$
$$\max(1 - f_{in}, 1 - f_{nj})]$$
$$= \min \max_k (1 - f_{ik}, 1 - f_{kj})$$

which is equal to the corresponding element of the matrix $\overline{F} * \overline{F}$. The proof for the n-matrix case follows similarly. As a matter of fact, the n matrices need not be the same.

Consider Z_{pq}, the complete set of $p \times q$ matrices with elements in Z. Two elements $S = [s_{ij}]$ and $T = [t_{ij}]$ in Z_{pq} are regarded equal iff $s_{ij} = t_{ij}$ for all i and j. We now define the following fuzzy compositions in Z_{pq}.

Definition 4.2.6.
(i) *Sum:* $S + T = W$ iff $w_{ij} = s_{ij} + t_{ij}$ for all i and j.
(ii) *Lattice product:* $S * T = W$ iff $w_{ij} = s_{ij}t_{ij}$ for all i and j.
(iii) $S \geqslant T$ iff $s_{ij} \geqslant t_{ij}$ for all i and j.
(iv) $S = \overline{T}$ iff $s_{ij} = \overline{t}_{ij}$ for all i and j.
Evidently, the system $[Z_{pq}, +, *]$ is again a lattice with the universal element E^* (all entries e^*) and the zero element E_+ (all entries e_+).

Definition 4.2.7.
Scalar multiplication: $W = rS = Sr$ iff $w_{ij} = rs_{ij}$, $S \in Z_{pq}$, $r \in Z$.
Matrix product: $W = ST$ iff $w_{ij} = \sum_{k=1}^{q} s_{ik}t_{kj}$, $S \in Z_{pq}$, $T \in Z_{qm}$, and $W \in Z_{pm}$, where the symbol $\sum$ is used to denote the sum with respect to the operation $+$ of Z.
Transpose: $W = S^t$ iff $w_{ij} = s_{ji}$, $S \in Z_{pq}$, $W \in Z_{qp}$.
Permanent:

$$\text{Per}(S) = |S|_{\text{per}} = \sum_i \left[\prod_j^p (s_{ji_j})\right],$$

where $S \in Z_{pp}$, $|S|_{\text{per}} \in Z$, and the summation is taken over all permutations $i_1, i_2, \ldots, i_p$.
Adjoint: $W = \text{Adj } S$ iff $w_{ij} = s_{ij}$ where $S \in Z_{pp}$, $W \in Z_{pp}$, and s_{ij} is the cofactor of s_{ji} in $|S|_{\text{per}}$.
It is interesting to note that Z_{pp} forms a lattice-ordered semi-group with the matrix product as the third binary composition, and the matrix I (all diagonal entries e^* and all others e_+) serves as unity.

Definition 4.2.8. A vector is called *fuzzy* iff all its entries are elements of Z. A matrix is called *fuzzy* iff all its rows are fuzzy vectors.

For the set of all $p \times p$ fuzzy matrices Z_{pp} we have the following.

Definition 4.2.9. Let $A = [a_{ij}]$ and $B = [b_{ij}]$ be two $p \times p$ fuzzy matrices:

(i) $A + B = [\text{Max}(a_{ij}, b_{ij})]$

(ii) $AB = [\underset{k}{\text{Max}}\{\min(a_{ik}, b_{kj})\}]$

(iii) $A \leq B$ exists iff $a_{ij} \leq b_{ij}, \forall\ i, j$.

It is easy to prove that $A \leq B$ implies $SA \leq SB$ and $AT \leq BT$, $\forall\ S, T \in Z_{pp}$.

Let A and B be two square matrices of order p and q respectively. The matrix

$$A \overset{\cdot}{+} B = \begin{bmatrix} A & 0 \\ 0 & B \end{bmatrix}$$

of order $p + q$ is called the *direct sum* and has the following properties:

(i) if A and B are fuzzy matrices, then so is $A \overset{\cdot}{+} B$;

(ii) $(A_1 \overset{\cdot}{+} B_1)(A_2 \overset{\cdot}{+} B_2) = A_1 A_2 \overset{\cdot}{+} B_1 B_2$ (provided the pairs A_1 and A_2, B_1 and B_2, are both of the same order). $A \leq B$ evidently implies $A + B = B$ and conversely. One immediately verifies that the set of $p \times p$ fuzzy matrices is a monoid under matrix multiplication. Namely, the set of $p \times p$ fuzzy matrices is closed under fuzzy matrix multiplication and the unit $p \times p$ matrix (all diagonal entries 1 and all others 0) is fuzzy.

Definition 4.2.10. A fuzzy matrix A is called *constant* if all its rows are equal.

The following propositions are immediate consequences of definitions and their proofs are quite trivial.

Proposition 4.2.1. If A is a constant fuzzy matrix and B a fuzzy matrix of the same order, then AB is a constant fuzzy matrix. If also $B \geq I$ then BA is constant.

Proposition 4.2.2. Let $A = [a_{ij}]$ be a constant fuzzy matrix and $B = [b_{ij}]$ a fuzzy matrix. If for all i, $\underset{j}{\text{Max}}[b_{ij}] \geq [a_{ij}]$, then $BA = A$.

Proposition 4.2.3. If A is a constant fuzzy matrix then $A^2 = A$ (i.e. fuzzy constant matrices are idempotent).

Proposition 4.2.4. Let A be a $p \times p$ symmetric fuzzy matrix. Then the sequence $A, A^2, A^3, \ldots$ is ultimately periodic.

Proposition 4.2.5. Let A be a $p \times p$ symmetric fuzzy matrix and let $B = A + I$. Then

$$B \leqslant B^2 \leqslant \ldots \leqslant B^{p-1} = B^p = B^{p+1} = \ldots$$

Let $\tilde{A} = \text{Sup}_k A^k$ where A is a $p \times p$ symmetric fuzzy matrix. From Proposition 4.2.5 it is clear that if $B = A + I$ then $\tilde{B} = B^{p-1}$.

Proposition 4.2.6. Let A be a $p \times p$ symmetric fuzzy matrix and let $B = A + I$ where I is of the same order. Then $\tilde{B} = \text{adj } B$.

Proposition 4.2.7. Let B be a $p \times p$ symmetric fuzzy matrix such that $B = A + I$. Then

(i) $\psi[\psi(B)] = \psi(B)$ $\qquad (\tilde{\tilde{B}} = \tilde{B})$

(ii) $B \leqslant \psi(B)$ $\qquad (\tilde{B} \geqslant B)$

(iii) $\psi(B^2) = \psi(B)$ $\qquad (\tilde{B}^2 = \tilde{B})$

Theorem 4.2.5. Let A and B be $p \times p$ symmetric fuzzy matrices and let I be a $p \times p$ unit matrix. If $C = A + I$ and $D = B + I$ then
$$\widetilde{C + D} = \widetilde{\tilde{C}\tilde{D}} = \widetilde{\tilde{D}\tilde{C}}.$$

Proof. $\tilde{C} \geqslant C$ and $\tilde{D} \geqslant I$ imply $\tilde{C}\tilde{D} \geqslant C$. Similarly $\tilde{C}\tilde{D} \geqslant D$ and therefore $\tilde{C}\tilde{D} \geqslant C + D$ which implies $\tilde{C}\tilde{D} \geqslant \widetilde{C + D}$. However, $\tilde{C} \leqslant \widetilde{C + D}$ and $\tilde{D} \leqslant \widetilde{C + D}$ imply $\tilde{C}\tilde{D} \leqslant \widetilde{C + D}$. The result follows immediately since
$$\widetilde{\tilde{C}\tilde{D}} \leqslant \widetilde{\widetilde{C + D}} = \widetilde{C + D}.$$ Q.E.D.

Theorem 4.2.6. Let A and B be $p \times p$ symmetric fuzzy matrices, and let $C = A + I$ and $D = B + I$, when I is the $p \times p$ unit matrix. Then

$C \leqslant \tilde{D}$ iff $\tilde{C} \leqslant \tilde{D}$.

Proof. $C \leqslant \tilde{C}$ implies that if $\tilde{C} \leqslant \tilde{D}$ then $C \leqslant \tilde{D}$.

$C \leqslant \tilde{D}$ implies $\tilde{C} \leqslant \tilde{\tilde{D}} = \tilde{D}$. Q.E.D.

Let S_k be a symmetric group on $1, 2, \ldots, k$, and let A be a fuzzy matrix of order k. The determinant of A, written $|A|$, is defined as

$$|A| = \sum_{\delta \in S_k} a_{1\delta(1)} a_{2\delta(2)} \ldots a_{k\delta(k)}.$$

It is obvious that $\min(|A|, |B|) = |\min(A,B)|$.

4.3. FUZZY AUTOMATA AND LANGUAGES

A fuzzy automaton was proposed by E.S. Santos[5] as a model of pattern recognition and automatic control systems. An advantage of employing a fuzzy automaton as a learning model is its simplicity in design and computation.

Let Σ be a finite non-empty alphabet. The set of all finite sequences over Σ is denoted by Σ^*. The null sequence is denoted by Λ and included in Σ^*; $|Q|$ is the number of elements in the set Q.

Definition 4.3.1. A finite fuzzy automaton over the alphabet Σ is a system

$$M = (Q, \pi, \{F(\sigma) | \sigma \in \Sigma\}, \eta^G)$$

where the following hold.
(1) $Q = \{q_1, q_2, \ldots, q_n\}$ is a non-empty finite set of internal states.
(2) π is an n-dimensional fuzzy row vector; that is,

$$\pi = (\pi_{q_1}, \pi_{q_2}, \ldots, \pi_{q_n})$$

where $0 \leq \pi_{q_i} \leq 1$, $1 \leq i \leq n$, and is called the *initial-state designator*.
(3) G is a subset of Q (the set of final states).
(4) $G = (\eta_{q_1}, \eta_{q_2}, \ldots, \eta_{q_n})'$ is an n-dimensional column vector, whose i^{th} component equals 1 if $q_i \in G$ and 0 otherwise, and is called the *final-state designator*.
(5) For each $\sigma \in \Sigma$, $F(\sigma)$ is a fuzzy matrix of order n (the fuzzy transition matrix of A) such that $F(\sigma) = [f_{q_i q_j}(\sigma)]$; $1 \leq i \leq n$, $1 \leq j \leq n$.

Let element $f_{q_i, q_j}(\sigma)$ of $F(\sigma)$ be $f(q_i, \sigma, q_j)$, where $q_i, q_j \in Q$ and $\sigma \in \Sigma$. The function f_A is a membership function of a fuzzy set in $Q \times \Sigma \times Q$; i.e.

$$f_A: Q \times \Sigma \times Q \to [0,1];$$

f_A may be called the *fuzzy transition function*. That is to say, for $s, t \in Q$, and $\sigma \in \Sigma$, $f_A(s, \sigma, t)$ is the grade of transition from state s to state t when the input is σ. The unity fuzzy transition function implies that such a transition may exist definitely.

If f_A takes only two values 0 and 1, then a fuzzy automaton A is a non-deterministic finite automaton. In addition, if only any one element of each row of matrix $F(\sigma)$, $\sigma \in \Sigma$ is 1, and the rest of the elements of each row are all equal to 0, then a fuzzy automaton A is a deterministic finite automaton.

The grade of transition for an input sequence of length m is defined by an m-ary fuzzy relation. The fuzzy transition function is as follows. For an input sequence $x = \sigma_1\sigma_2\ldots\sigma_m \in \Sigma_m^*$ and $s,t \in Q$

$$f_M(s,x,t) = \max_{q_1,q_2,\ldots,q_{m-1}\in Q} \min f_A(s,\sigma_1,q_1), f_A(q_1,\sigma_2,q_2),\ldots,$$
$$f_A(q_{m-1},\sigma_m,t)$$

which is the grade of transition from state s to state t when the input sequence is $x = \sigma_1\sigma_2\ldots\sigma_m$.

Definition 4.3.2. For Λ, x, $y \in \Sigma^*$ and s, $t \in Q$,

$$f_A(s,\Lambda,t) = \begin{cases} 1 & \text{if } s = t \\ 0 & \text{if } s \neq t, \end{cases}$$

$$f_A(s,xy,t) = \max_{q\in Q} \min f_A(s,x,q), f_A(q,y,t).$$

In particular we call a fuzzy automaton with the grade of transition under the operation max min a *pessimistic fuzzy automaton* (p.f.a.), and a fuzzy automaton under the operation min max an *optimistic fuzzy automaton* (o.f.a.).

Definition 4.3.3. An optimistic fuzzy automaton over the alphabet M' is a system

$$M' = (Q',\pi',\{F'(\sigma)\,|\,\sigma \in \Sigma'\},\eta^{G'})$$

where Q' is a finite non-empty set of the internal states of M' and $|Q'| = n'$. π' is an n'-dimensional row vector (the initial state designator). A *fuzzy transition function* $f_{M'}'$ is defined as follows. For Λ, x, $y \in \Sigma'^*$ and s, $t \in Q'$,

$$f_{M'}'(s,\Lambda,t) = \begin{cases} 0 & \text{if } s = t \\ 1 & \text{if } s \neq t \end{cases}$$

$$f_{M'}'(s,xy,t) = \min_{q\in Q'} \max \left[f_{M'}'(s,x,q), f_{M'}'(q,y,t)\right].$$

G' is a subset of Q' (the set of final states), and an n'-dimensional column vector (the final-state designator)

$$\eta^{G'} = (\eta_{q_1},\eta_{q_2},\ldots,\eta_{q_{n'}})$$

is defined such that $\eta_{q_i}' = 0$ if $q_i \in G'$ and $\eta_{q_i} = 1$ otherwise.

Note that an element of zero in π' means the definite existence of such an initial state. Unless stated to the contrary, by a "fuzzy automaton" we shall mean a pessimistic automaton.

E. S. Santos[5] showed that the capability of fuzzy automata as recognizers is the same as that of finite automata, though fuzzy automata include the deterministic and non-deterministic finite automata as special cases.

Definition 4.3.4. Let $M = (Q, \pi, \{F(\sigma) | \sigma \in \Sigma\}, \eta^G)$ be a fuzzy automaton and λ a real number $0 \leqslant \lambda < 1$. The set of all input sequences accepted by A with parameter λ is defined as

$$L(M, \circ, \lambda) = \{x | f_M(x) > \lambda, x \in \Sigma^*\}$$

where λ is called a *threshold* of A and $L(M, \circ, \lambda)$ a λ-*fuzzy language.*

Theorem 4.3.1. (Santos). A λ-fuzzy language $L(M, \circ, \lambda)$ is a regular language.

The same theorem also holds for an optimistic fuzzy automaton.

Definition 4.3.5. For a fuzzy matrix $A = [a_{ij}]$, $0 \leqslant a_{ij} \leqslant 1$ and a real number d such that $0 \leqslant d \leqslant 1$, we define a fuzzy matrix $A' = [a'_{ij}]$ as follows:

$$a'_{ij} = \begin{cases} a_{ij} + d & \text{if } a_{ij} \leqslant 1 - d \\ 1 & \text{otherwise} \end{cases}$$

Lemma 4.3.1. For two fuzzy matrices U and V of the same order, let the fuzzy matrices defined in Definition 4.3.5 be U' and V', respectively; then, for two fuzzy matrices $W = [w_{ij}]$ and $W' = [w'_{ij}]$ such that $W = U \circ V$ and $W' = U' \circ V'$, we have that

$$w'_{ij} = \begin{cases} w_{ij} + d & \text{if } w_{ij} \leqslant 1 - d \\ 1 & \text{otherwise} \end{cases}$$

Proof. The proof is clear from the property of the operation $\circ$.

Definition 4.3.6. For a fuzzy matrix $A = [a_{ij}]$ and d' a real number, $0 \leqslant d' \leqslant 1$, define a fuzzy matrix $A'' = [a''_{ij}]$ as follows:

$$a''_{ij} = \begin{cases} a_{ij} - d' & \text{if } a_{ij} \geqslant d' \\ 0 & \text{otherwise.} \end{cases}$$

Lemma 4.3.2. For two fuzzy matrices U and V of the same order, let the fuzzy matrices defined in Definition 4.3.6 be U'' and V'', respectively; then, for two fuzzy matrices $W = [w_{ij}]$ and $W' = [w'_{ij}]$ such that $W = U \circ V$ and $W'' = U'' \circ V'',$ we have that

$$w''_{ij} = \begin{cases} w_{ij} - d' & \text{if } a_{ij} \geqslant d' \\ 0 & \text{otherwise} \end{cases}$$

Proof. Again the proof of this lemma follows directly from the property of the operation ∘.

Theorem 4.3.2. Every fuzzy language is λ-fuzzy for any λ such that $0 \leq \lambda \leq 1$.

Proof. Let

$$L = L(M, \circ, \mu)$$

and let

$$M = \left(Q, \pi, \{F(\sigma) \,|\, \sigma \in \Sigma\}, \eta^G\right)$$

be a fuzzy automaton, where

$$F(\sigma) = \left[f_{q_i,q_j}(\sigma)\right], \pi = (\pi_{q_i}), q_i, q_j \in Q, \text{ and } \sigma \in \Sigma.$$

Omitting the trivial case $\lambda = \mu$, we can assume that $\lambda \neq \mu$.

Case (1) $\lambda > \mu$. Consider the fuzzy automaton

$$M' = \left(Q', \pi', \{F'(\sigma) \,|\, \sigma \in \Sigma'\}, \eta^{G'}\right)$$

where $Q' = Q$, $\Sigma' = \Sigma$, $\eta^{G'} = \eta^G$, and the transition matrices

$$F'(\sigma) = \left[f'_{q_i,q_j}(\sigma)\right], \sigma \in \Sigma'$$

and

$$\pi' = (\pi'_{q_i})$$

are defined as follows:

$$f'_{q_i,q_j}(\sigma) = \begin{cases} f_{q_i,q_j}(\sigma) + (\lambda - \mu) & \text{if } f_{q_i,q_j}(\sigma) \leq 1 - \lambda + \mu \\ 1 & \text{otherwise} \end{cases}$$

$$\pi'_{q_i} = \begin{cases} \pi_{q_i} + (\lambda - \mu) & \text{if } \pi_{q_i} \leq 1 - \lambda + \mu \\ 1 & \text{otherwise.} \end{cases}$$

Thus, according to Lemma 4.3.1, for $x \in \Sigma^*$

$$\pi' \circ F'(x) \circ \eta^{G'} = \begin{cases} \pi \circ F(x) \circ \eta^G + (\lambda - \mu) & \text{if } \pi \circ F(x) \circ \eta^G \leq 1 - \lambda + \mu \\ 1 & \text{otherwise.} \end{cases}$$

Therefore, $L(M', \circ, \lambda) = L(M, \circ, \mu)$ when $\lambda > \mu$.

Case (2) $\lambda < \mu$. Consider the fuzzy automaton

$$M'' = \left(Q'', \pi'', \{F''(\sigma) \,|\, \sigma \in \Sigma''\}, \eta^{G''}\right)$$

where $Q'' = Q$, $\Sigma'' = \Sigma$, $\eta^{G''} = \eta^G$, and the fuzzy transition matrices

$$F''(\sigma) = \left[f''_{q_i q_j}(\sigma)\right]$$

where $\sigma \in \Sigma''$ and $\pi'' = (\pi''_{q_i})$ are defined as follows:

$$f''_{q_i q_j}(\sigma) = \begin{cases} f_{q_i q_j}(\sigma) - (\mu - \lambda) & \text{if } f_{q_i q_j}(\sigma) > \mu - \lambda \\ 0 & \text{otherwise} \end{cases}$$

$$\pi''_{q_i} = \begin{cases} \pi_{q_i} - (\mu - \lambda) & \text{if } \pi_{q_i} \geqslant \mu - \lambda \\ 0 & \text{otherwise.} \end{cases}$$

Thus, according to Lemma 4.3.2, for $x \in \Sigma^*$

$$\pi'' \circ F''(x) \circ \eta^{G'} = \begin{cases} \pi \circ F(x) \circ \eta^G - (\mu - \lambda) \\ \qquad \text{if } \pi \circ F(x) \circ \eta^G \geqslant \mu - \lambda \\ 0 \qquad \text{otherwise.} \end{cases}$$

Therefore, $L(M'', \circ, \lambda) = L(M, \circ, \mu)$ when $\lambda < \mu$. Hence, it follows that in both cases $L = L(M', \circ, \lambda)$ or $L(M'', \circ, \lambda)$, which implies our theorem.

It is easily shown that the same theorem holds for an optimistic fuzzy automaton.

In the following, we use the concept of fuzzy sets instead of the set of input sequences with threshold λ and show that a family of fuzzy events characterized by not only pessimistic fuzzy automata (p.f.a.) but also optimistic fuzzy automata (o.f.a.) is closed under the operations of intersection and union in the fuzzy sense, and the complement of the fuzzy event by a p.f.a. (an o.f.a.) is characterized by an o.f.a. (a p.f.a.).

Definition 4.3.7. For a p.f.a.

$$M = \left(Q, \pi, \{F(\sigma) \,|\, \sigma \in \Sigma\}, \eta^G\right)$$

let a fuzzy event be the fuzzy set in Σ^* which is characterized by

$$f_M(x) = \pi \circ F(x) \circ \eta^G, \text{ where } x \in \Sigma^*.$$

We denote by $\tilde{L}(M, \circ)$ the fuzzy event by a p.f.a. M and, similarly, by $\tilde{L}(M', *)$ the fuzzy event by an o.f.a. M'.

Definition 4.3.8. For two p.f.a.'s M_1 and M_2

$$M_1 = \left(Q_1, \pi_1, \{F_1(\sigma) \,|\, \sigma \in \Sigma\}, \eta^{G_1}\right)$$

$$M_2 = \left(Q_2, \pi_2, \{F_2(\sigma) \,|\, \sigma \in \Sigma\}, \eta^{G_2}\right)$$

define a min p.f.a. $M_1 \otimes M_2$ as follows:

$$M_1 \otimes M_2 = (Q, \pi, \{F(\sigma) \mid \sigma \in \Sigma\}, \eta^G)$$

where

$$Q = Q_1 \times Q_2 = \{(s_i, t_j) \; s_i \in Q_1, t_j \in Q_2, 1 \leq i \leq m, 1 \leq j \leq n\}$$

$G = G_1 \times G_2$, $m = |Q_1|$ and $n = |Q_2|$. As to the fuzzy transition function $f_{M_1 \otimes M_2}$ of a min p.f.a. $M_1 \otimes M_2$, define

$$f_{M_1 \otimes M_2}[(s,t), \sigma, (q,r)] = \min[f_{M_1}(s, \sigma, q), f_{M_2}(t, \sigma, r)]$$

for $(s,t), (q,r) \in Q$ and $\sigma \in \Sigma$.

Moreover, the mn-dimensional row vector π is defined as follows. For $(s_i, t_j) \in Q$, $1 \leq i \leq m$ and $1 \leq j \leq n$,

$$\pi = \pi_1 \otimes \pi_2 = (\xi_{(s_i, t_j)})$$

where

$$\xi_{(s_i, t_j)} = \min[\pi_{1 s_i}, \; \pi_{2 t_j}]$$

and

$$(s_i, t_j) = (s_1, t_1), (s_1, t_2), \ldots, (s_1, t_n), (s_2, t_1), \ldots, (s_m, t_n).$$

The mn-dimensional column vector η^G is also $\eta^G = \eta^{G_1} \otimes \eta^{G_2}$.

Hence, the fuzzy transition matrices of order mn of $M_1 \otimes M_2$ is as follows. For two p.f.a.'s M_1 and M_2, let

$$F_1(\sigma) = [f_{s_i, s_j}(\sigma)]$$

and

$$F_2(\sigma) = [f_{t_k, t_l}(\sigma)],$$

where $\sigma \in \Sigma$, be fuzzy transition matrices of M_1 and M_2, respectively; then the fuzzy transition matrices $F(\sigma)$, $\sigma \in \Sigma$ of $M_1 \otimes M_2$ are defined by

$$F(\sigma) = F_1(\sigma) \otimes F_2(\sigma) = [f_{(s_i, t_k), (s_j, t_l)}(\sigma)]$$

where

$$f_{(s_i, t_k), (s_j, t_l)}(\sigma) = \min[f_{s_i, s_j}(\sigma), f_{t_k, t_l}(\sigma)]$$
$$= f_{M_1 \otimes M_2}[(s_i, t_k), \sigma, (s_j, t_l)].$$

Note that the operation $\otimes$ of fuzzy matrices corresponds to the tensor product of ordinary matrices.

Lemma 4.3.3. For fuzzy matrices $A_1, A_2, B_1, B_2, A,$ and $B,$ for row vectors π_1 and π_2, and for column vectors η^{G_1} and η^{G_2} we have that

(1) $(A_1 \circ B_1) \otimes (A_2 \circ B_2) \qquad = (A_1 \otimes A_2) \circ (B_1 \otimes B_2)$

(2) $(\pi_1 \circ A \circ \eta^{G_1}) \otimes (\pi_2 \circ B \circ \eta^{G_2}) = (\pi_1 \otimes \pi_2) \circ (A \otimes B)$
$$\circ (\eta^{G_1} \otimes \eta^{G_2})$$
$$= \min[\pi_1 \circ A \circ \eta^{G_1}, \ \pi_2 \circ B \circ \eta^{G_2}]$$

(3) $A_1 \otimes A_2, \pi_1 \otimes \pi_2, \ldots,$ *are fuzzy matrices.*

Proof. The proof of this lemma is rather straightforward and is left to the reader.

This lemma enables us to prove the following closure theorem.

Theorem 4.3.3. Let $M_1, M_2,$ and $M_1 \otimes M_2$ be p.f.a.'s as in Definition 4.3.8 and $\tilde{L}(M_1, \circ),$ $\tilde{L}(M_2, \circ)$ and $L(M_1, M_2, \circ)$ be the fuzzy events characterized by $M_1, M_2,$ and $M_1 \otimes M_2,$ respectively. Then, in the fuzzy sense,

$$\tilde{L}(M_1, \circ) \cap \tilde{L}(M_2, \circ) = \tilde{L}(M_1 \otimes M_2, \circ)$$

Proof. The membership functions of fuzzy events $\tilde{L}(M_1, \circ),$ $\tilde{L}(M_2, \circ)$ and $\tilde{L}(M_1 \otimes M_2, \circ)$ are

$$f_{M_1}(x) = \pi_1 \circ F_1(x) \circ \eta^{G_1}$$
$$f_{M_2}(x) = \pi_2 \circ F_2(x) \circ \eta^{G_2}$$

and

$$f_{M_1 \otimes M_2}(x) = \pi \circ F(x) \circ \eta^G$$

respectively, where $x \in \Sigma^*$. By Lemma 4.3.3, for $x \in \Sigma^*$,

$$f_{M_1 \otimes M_2}(x) = \pi \circ F(x) \circ \eta^G$$
$$= (\pi_1 \otimes \pi_2) \circ (F_1(x) \otimes F_2(x)) \circ (\eta^{G_1} \otimes \eta^{G_2})$$
$$= \min[\pi_1 \circ F_1(x) \circ \eta^{G_1}, \ \pi_2 \circ F_2(x) \circ \eta^{G_2}]$$
$$= \min[f_{M_1}(x), f_{M_2}(x)].$$

Corollary 4.3.1. For two o.f.a.'s M_1' and M_2', let $\tilde{L}(M_1', *)$ and $\tilde{L}(M_2', *)$ be the fuzzy events by M_1' and M_2', respectively; then, in the fuzzy sense, there exists an o.f.a. M such that

$$\tilde{L}(M_1', *) \cup \tilde{L}(M'_2, *) = L(M, *).$$

This corollary can be proven by replacing the operation min by the operation max in Definition 4.3.8 and defining a max o.f.a.

We have shown that the family of fuzzy events by p.f.a. is closed under intersection and the family by o.f.a. is closed under union in the fuzzy sense. Next, we shall verify that the family of fuzzy events by p.f.a. (o.f.a.) is closed under union (intersection) in the fuzzy sense.

Theorem 4.3.4. For two p.f.a.'s M_1 and M_2, let $\tilde{L}(M_1, \circ)$ and $\tilde{L}(M_2, \circ)$ be fuzzy events by M_1 and M_2, respectively; then, in the fuzzy sense, there exists a p.f.a. M such that

$$\tilde{L}(M_1, \circ) \cup \tilde{L}(M_2, \circ) = \tilde{L}(M, \circ).$$

Proof. Let M_1 and M_2 be two p.f.a.'s as follows:

$$M_1 = (\{s_1, s_2, \ldots, s_m\}, \pi_1, \{F_1(\sigma) \mid \sigma \in \Sigma\}, \eta^{G_1})$$

$$M_2 = (\{t_1, t_2, \ldots, t_n\}, \pi_2, \{F_2(\sigma) \mid \sigma \in \Sigma\}, \eta^{G_2}).$$

Now, consider a p.f.a. M; that is

$$M = (\{s_1, \ldots, s_m, t_1, \ldots, t_n\}, \pi, \{F(\sigma) \mid \sigma \in \Sigma\}, \eta^G)$$

where π, $F(\sigma)$ and η^G are given as follows. If

$$\pi_1 = (\pi_{s_1}, \pi_{s_2}, \ldots, \pi_{s_m}) \quad \text{and} \quad \pi_2 = (\pi_{t_1}, \pi_{t_2}, \ldots, \pi_{t_n})$$

then

$$\pi = (\pi_{s_1}, \ldots, \pi_{s_m}, \pi_{t_1}, \ldots, \pi_{t_n}) = (\pi_1 \pi_2).$$

Moreover,

$$F(\sigma) = \begin{bmatrix} F_1(\sigma) & 0 \\ 0 & F_2(\sigma) \end{bmatrix}$$

and

$$\eta^G = \begin{bmatrix} \eta^{G_1} \\ \eta^{G_2} \end{bmatrix}.$$

In general, in fuzzy matrices, we have that

$$(1) \quad \begin{bmatrix} A_1 & 0 \\ 0 & B_1 \end{bmatrix} \circ \begin{bmatrix} A_2 & 0 \\ 0 & B_2 \end{bmatrix} = \begin{bmatrix} A_1 \circ A_2 & 0 \\ 0 & B_1 \circ B_2 \end{bmatrix}$$

$$(2) \quad \begin{bmatrix} \pi_1 & \pi_2 \end{bmatrix} \circ \begin{bmatrix} A & 0 \\ 0 & B \end{bmatrix} \circ \begin{bmatrix} \eta^{G_1} \\ \eta^{G_2} \end{bmatrix} = \max \begin{bmatrix} \pi_1 \circ A \circ \eta^{G_1}, \pi_2 \circ B \circ \eta^{G_2} \end{bmatrix}$$

Therefore, let

$$f_{A_1}(x) = \pi_1 \circ F_1(x) \circ \eta^{G_1}$$

$$f_{A_2}(x) = \pi_2 \circ F_2(x) \circ \eta^{G_2}$$

and

$$f_A(x) = \pi \circ F(x) \circ \eta^G$$

where $x \in \Sigma^*$ are the membership functions which characterize fuzzy events $\tilde{L}(A_1, \circ)$, $\tilde{L}(A_2, \circ)$, and $\tilde{L}(A, \circ)$, respectively. Then, for $x \in \Sigma^*$, we have

$$f_A(x) = \pi \circ F(x) \circ \eta^G = \begin{bmatrix} \pi_1 & \pi_2 \end{bmatrix} \circ \begin{bmatrix} F_1(x) & 0 \\ 0 & F_2(x) \end{bmatrix} \circ \begin{bmatrix} \eta^{G_1} \\ \eta^{G_2} \end{bmatrix}$$

$$= \max \begin{bmatrix} \pi_1 \circ F_1(x) \circ \eta^{G_1}, \pi_2 \circ F_2(x) \circ \eta^{G_2} \end{bmatrix}$$

$$= \max \begin{bmatrix} f_{A_1}(x), f_{A_2}(x) \end{bmatrix}$$

Corollary 4.3.2. For two o.f.a.'s M_1' and M_2', let $\tilde{L}(M_1', *)$ and $\tilde{L}(M_2', *)$ be the fuzzy events M_1' and M_2', respectively; then, in the fuzzy sense, there exists an o.f.a. M such that

$$\tilde{L}(M_1', *) \cap \tilde{L}(M_2', *) = *L(M, *)$$

Proof. For two o.f.a.'s M_1' and M_2', that is

$$M_1' = (Q_1, \pi_1, \{F_1(\sigma) | \sigma \in \Sigma\}, \eta^{G_1})$$

$$M_2' = (Q_2, \pi_2, \{F_2(\sigma) | \sigma \in \Sigma\}, \eta^{G_2})$$

let us define an o.f.a. M' as follows:

$$M' = (Q, \pi, \{F(\sigma) | \sigma \in \Sigma\}, \eta^G)$$

where $Q = Q_1 \cup Q_2$, $Q_1 \cap Q_2 = \varnothing$, $\pi = \begin{bmatrix} \pi_1 & \pi_2 \end{bmatrix}$

$$F(\sigma) = \begin{bmatrix} F_1(\sigma) & 1 \\ 1 & F_2(\sigma) \end{bmatrix} \qquad \text{for all } \sigma \text{ in } \Sigma$$

$$\eta^G = \begin{bmatrix} \eta^{G_1} \\ \eta^{G_2} \end{bmatrix}$$

Then, we can prove our corollary immediately in a similar way as for Theorem 4.3.4. We shall show the inclusion property of p.f.a.'s.

Theorem 4.3.5. Given two p.f.a.'s M_1 and M_2 as follows:

$$M_1 = (Q_1, \pi_1, \{F_1(\sigma) \,|\, \sigma \in \Sigma\}, \eta^{G_1})$$

$$M_2 = (Q_2, \pi_2, \{F_2(\sigma) \,|\, \sigma \in \Sigma\}, \eta^{G_2})$$

If $|Q_1| = |Q_2|$, $F_1(\sigma) < F_2(\sigma)$ *for all* σ *in* Σ, $\pi_1 < \pi_2$, *and* $\eta^{G_1} < \eta^{G_2}$, *then, in the fuzzy sense,*

$$\tilde{L}(M_1, \circ) \subseteq \tilde{L}(M_2, \circ).$$

Proof. We can easily show that

$$f_{M_1}(x) \leqslant f_{M_2}(x) \quad \text{for } x \in \Sigma^*$$

from the basic properties of fuzzy matrices described in §4.2. Obviously, the same theorem also holds for o.f.a.'s.

We shall show that the complement of a fuzzy event by a p.f.a. (an o.f.a.) is characterized by an o.f.a. (a p.f.a.).

Definition 4.3.9. If $M = (Q, \pi, \{F(\sigma) \,|\, \sigma \in \Sigma\}, \eta^G)$ is a p.f.a., the *complementary o.f.a. for A* is defined as

$$\overline{M} = (\overline{Q}, \overline{\pi}, \{\overline{F}(\sigma) \,|\, \sigma \in \Sigma\}, \eta^M)$$

where $\overline{Q} = Q$. As to the fuzzy transition function $\overline{f_M}$ of $\overline{M}$, for $\sigma \in \Sigma$, $x, y \in \Sigma^*$, $s, t \in Q$, and f_M of M, we define

$$\overline{f_M}(s, \sigma, t) = 1 - f_M(s, \sigma, t)$$

$$\overline{f_M}(s, xy, t) = \min \max \left[\overline{f_Q}(s, x, l), \overline{f_Q}(l, y, t) \right]$$

$$= 1 - f_M(s, xy, t)$$

and the initial and final state vectors are

$$\overline{\pi} = [1, 1, \dots, 1] - \pi, \quad \text{and} \quad \eta^{\overline{G}} = [1, 1, \dots, 1]' - \eta^G.$$

Note that we can easily define a complementary p.f.a. M' for an o.f.a. M' in a similar way.

Lemma 4.3.4. For a fuzzy matrix $U = [u_{ij}]$, let $U' = [u'_{ij}]$ be a fuzzy matrix such that

$$u'_{ij} = 1 - u_{ij}.$$

For fuzzy matrices $U_1, U_2, \dots, U_m$, let $U'_1, U'_2, \dots, U'_m$ be fuzzy matrices as defined above, respectively; then

$$U_1 \circ U_2 \circ \dots \circ U_m + U'_1 * U'_2 * \dots * U'_m = \begin{bmatrix} 1 \dots 1 \\ \vdots \quad \vdots \\ 1 \dots 1 \end{bmatrix}$$

Theorem 4.3.6. Let M be a p.f.a. and let $\overline{M}$ be a complementary o.f.a. for M; then, in the fuzzy sense,

$$\overline{\tilde{L}(M, \circ)} = \tilde{L}(\overline{M}, *).$$

Proof. Let $M = (Q, \pi, \{F(\sigma) \,|\, \sigma \in \Sigma\}, \eta^G)$ be a p.f.a. and $\overline{M} = (Q, \overline{\pi}, \{\overline{F}(\sigma) \,|\, \sigma \in \Sigma\}, \eta^{\overline{G}})$ be a complementary o.f.a. for M, then by Lemma 4.3.4, for $x \in \Sigma^*$,

$$f_M(x) = \pi \circ F(x) \circ \eta^G = 1 - \overline{\pi} * \overline{F}(x) * \eta^{\overline{G}} = 1 - \overline{f_{\overline{M}}}(x).$$

Therefore, we have $\overline{\tilde{L}(M, \circ)} = \tilde{L}(\tilde{M}, *)$.

Corollary 4.3.3. For an o.f.a. M' and a complementary p.f.a. $\overline{M}'$ for M', in the fuzzy sense,

$$\overline{\tilde{L}(M', *)} = \tilde{L}(\overline{M'}, \circ).$$

Proof. Immediately.

The family of fuzzy events characterized by a p.f.a. (o.f.a.) constitutes a distributive lattice, but does not constitute a Boolean lattice clearly.

The threshold of fuzzy automata can be set arbitrarily by changing the value of each element of the fuzzy transition matrix and the initial-state designator. Moreover, a family of fuzzy events characterized by pessimistic (optimistic) fuzzy automata is closed under the operations of union and intersection in the sense of fuzzy sets. The complement of the fuzzy event by a pessimistic (optimistic) fuzzy automaton is characterized by an optimistic (pessimistic) fuzzy automaton.

4.4. REGULAR EXPRESSIONS AND REGULAR FUZZY LANGUAGES

Given a REGULAR FUZZY GRAMMAR (RFG) or FUZZY FINITE AUTOMATON (FFA), it is known that a non-fuzzy "threshold language"

$$L(\theta) = \{x \colon x \in \Sigma^*, \ \mu_{\mathrm{RFL}}(x) \geq \theta\}, \qquad 0 < \theta \leq 1$$

is regular where RFL stands for REGULAR FUZZY LANGUAGE and there exists a regular expression $F'(\theta)$ describing $L(\theta)$. Given regular expressions $F'(\theta_1)$ and $F'(\theta_2)$, the Boolean functions of $F'(\theta_1)$ and $F'(\theta_2)$ are also regular expressions[9]. Specifically, if $\theta_1 > \theta_2$, then

$$F'(\theta_2) \cap \overline{F'(\theta_1)}$$

where $\overline{F'(\theta_1)}$ denotes the complement of $F'(\theta_1)$, is a regular expression which describes those strings in $L(\theta_2)$ and not in $L(\theta_1)$.

Consider the finite set M of possible non-zero membership grades in an RFL. Choose two adjacent membership grades θ_1, θ_2 in M such that $\theta_1 > \theta_2$. Then the regular expression

$$F(\theta_2) = F'(\theta_2) \cap \overline{F'(\theta_1)}$$

defines the set of strings

$$\{x : x \in \Sigma^*, \mu_{RFL}(x) = \theta_2\}.$$

$F(\theta_2)$ defines an equivalence class of an equivalence relation R in Σ^* where x is R-related to y iff $\mu_{RFL}(x) = \mu_{RFL}(y)$.

The regular expressions $F(m)$, $M \in M$, define the state-transition diagram of a non-fuzzy, deterministic finite automaton DFA $= (\Sigma, Q, \delta, S_0, T)$ with an output map $\omega: Q \to M \cup \{0\}$, where

$$T = \{F : F \in Q, \ \omega(F) = m, \ m \in M\}.$$

Thus, the set of strings which have membership grade m in the RFL is identically the set of strings which cause the DFA, started in state S_0, to halt in the state with output m. From this DFA we obtain an RFG which generates the RFL and has productions in a normal form.

Theorem 4.4.1. *Let RFG_1 be a regular fuzzy grammar for which the max(min) derivation rule is used. Then there is an equivalent $RFG_2 = (V_N, V_T, S_0, P)$ in which productions have the form $A \xrightarrow{1.0} aB$ or $A \xrightarrow{\sigma} a$ where $A, B \in V_N$, $a \in V_T$, and $0 < \sigma \leq 1.0$.*

Proof. The proof is a six-step algorithm for constructing RFG_2 from RFG_1.

Step 1: Given RFG_1, obtain the corresponding FFA.
Remark: The FFA is obtained from RFG_1 in the same way that a non-fuzzy FUZZY AUTOMATON (FA) is obtained from a non-fuzzy REGULAR GRAMMAR (RG) (cf. Theorem 3.4 in [10]) with the exception that a production membership grade is assigned to the corresponding transition of the FFA.

Step 2: Obtain the set M of possible membership grades.
Remark: Initially, each production membership grade is a possible membership grade of a string in the RFL; therefore, M may be taken to be the finite set of distinct production membership grades.

Step 3: For each $m \in M$, obtain the regular expression $F'(m)$ describing those strings $x \in \Sigma^*$ such that $\mu_{\mathrm{RFL}}(x) \geqslant m$.

Remark: Consider a "sliding threshold" θ, $0 < \theta \leqslant 1$, which is in turn set equal to each element of M from lowest to highest. For each value of θ obtain a non-fuzzy, non-deterministic FA from FFA by retaining (without membership grade) only those transitions with membership grade greater than or equal to θ. Convert this NON-DETERMINATE FINITE AUTOMATON (NDFA) to a deterministic FA (cf. Theorem 3.3 in [10]) and find the regular expression $F'(\theta)$ by standard techniques [9].

Step 4: For each m in M, obtain the regular expression $F(m)$ describing those strings $x \in \Sigma^*$ such that $\mu_{\mathrm{RFL}}(x) = m$.

Remark: Beginning with the lowest membership grade in M, compare each pair of regular expressions $F'(m)$, $F'(n)$ where m,n in M are adjacent and $m > n$. If $F'(n) \cap F'(m) \neq \varnothing$, set $F(n)$ equal to the new regular expression

$$F'(n) \cap \overline{F'(m)}.$$

If $F'(n) \cap F'(m) = \varnothing$, then set $F(n)$ equal to $F'(n)$. This procedure makes all regular expressions $F(m)$, $m \in M$, pair-wise disjoint.

Step 5: Use the regular expressions $F(m)$, $m \in M$, to obtain the state-transition diagram of the DFA.

Remark: The procedure for obtaining a state-transition diagram by taking "derivatives" of a regular expression is well documented[9,11].

Step 6: Obtain the normal form productions for RFG_2 from the transition diagram of the DFA.

Remark: There is a production $A \xrightarrow{0.1} aB$ (or $A \rightarrow aB$, with 1.0 understood) in P if there is a transition $\delta(A,a) = B$; there is a production $A \xrightarrow{m} a$ in P if there is a transition $\delta(A,a) = F$, where $\omega(F) = m \neq 0$. The initial symbol of RFG_2 corresponds to the initial state S_0 of the DFA.

The construction of the DFA is such that the set of strings which cause the DFA to halt in a state with output m is identically the set of strings generated by RFG_1 with membership grade m as determined by the max(min) rule. Suppose the string $ab \ldots g$ causes the DFA to enter sequentially states $S_0, A, B, \ldots, F$ where $\omega(F) = m$; then there is a derivation

$$S \rightarrow aA \rightarrow abB \rightarrow \ldots \xrightarrow{m} ab \ldots g$$

in RFG_2. On the other hand, such a derivation in RFG_2 yields a string $ab \ldots g$ which causes the DFA, started in state S_0, to halt in state F with $\omega(F) = m$.

We conclude that RFG_2 is equivalent to RFG_1 and has all productions in the required normal form. Since it is obtained from a deterministic FA, RFG_2 is unambiguous.

Illustration of the Algorithm. Consider $RFG_1 = (V_N, V_T, S_0, P)$ where $V_N = \{S_0, A, B\}$, $V_T = \{a, b\}$, and the productions are as follows:

$$S_0 \xrightarrow{0.3} aS_0, \quad S_0 \xrightarrow{0.5} aA, \quad S_0 \xrightarrow{0.7} aB$$

$$S_0 \xrightarrow{0.3} bS_0, \quad S_0 \xrightarrow{0.3} bA, \quad A \xrightarrow{0.5} b, \quad B \xrightarrow{0.4} b.$$

Step 1: The corresponding FFA is shown in Fig. 4.2.

Step 2: $M = \{0.7, 0.5, 0.4, 0.3, 0.2\}$.

Step 3: $\theta = 0.2, \quad F'(0.2) = (a + b) * (ab + bb)$

$\qquad \theta = 0.3, \quad F'(0.3) = (a + b) * ab$

$\qquad \theta = 0.4, \quad F'(0.4) = ab$

$\qquad \theta = 0.5, \quad F'(0.5) = ab$

$\qquad \theta = 0.7, \quad F'(0.7) = \varnothing.$

For this FFA, the above regular expressions are written by inspection.

Step 4: $F'(0.2) \cap F'(0.3) = (a + b) * ab$

$\qquad\qquad F(0.2) = F'(0.2) \cap F'(0.3) = (a + b) * bb$

$\quad F'(0.3) \cap F'(0.4) = ab$

$\qquad\qquad F(0.3) = F'(0.3) \cap F'(0.4) = (a + b)(a + b) * ab$

$\qquad\qquad\qquad\qquad\qquad\qquad = (a + b) * ab$

$\quad F'(0.4) \cap F'(0.5) = ab$

$\qquad\qquad F(0.4) = F'(0.4) \cap F'(0.5) = \varnothing$

$\quad F'(0.5) \cap F'(0.7) = \varnothing$

$\qquad\qquad F(0.5) = F'(0.5) = ab$

$\qquad\qquad F(0.7) = F'(0.7) = \varnothing.$

This asserts, for instance, that $\mu_{\text{RFL}}(ababb) = 0.2$ since string $ababb$ is contained in the set described by $F(0.2)$.

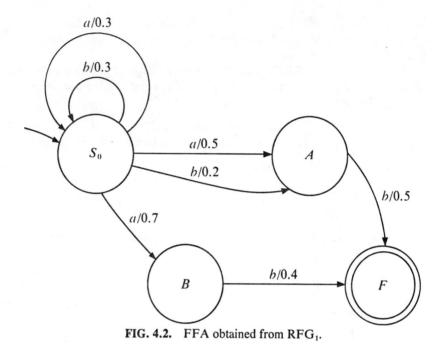

FIG. 4.2.   FFA obtained from RFG$_1$.

Step 5: The distinct derivatives of $F = \left[F(0.5), F(0.3), F(0.2)\right]$ are given in Table 4.2. The transition diagram of the DFA is shown in Fig. 4.3.

Step 6: The productions in RFG$_2$ are

$$S_0 \rightarrow aA \mid bB$$

$$A \rightarrow aC \mid bD \qquad A \xrightarrow{0.5} b$$

$$B \rightarrow aC \mid bE \qquad B \xrightarrow{0.2} b$$

$$C \rightarrow aC \mid bF \qquad C \xrightarrow{0.3} b$$

$$D \rightarrow aC \mid bE \qquad D \xrightarrow{0.3} b$$

$$E \rightarrow aC \mid bE \qquad E \xrightarrow{0.3} b$$

$$F \rightarrow aC \mid bE \qquad F \xrightarrow{0.3} b.$$

Example 4.4.1. Consider string *ababb*. Using RFG$_1$,

$$S_0 \xrightarrow{0.3} aS_0 \xrightarrow{0.3} abS_0 \xrightarrow{0.3} abaS_0 \xrightarrow{0.3} ababA \xrightarrow{0.5} ababb$$

Derivative	State in Q	δ	ω
$D_\lambda F = [ab, (a+b)*ab, (a+b)*bb]$	S_0	$\delta(S_0,a) = A$	0
$D_a F = [b, (a+b)*ab, (a+b)*bb]$	A	$\delta(S_0,b) = B$	0
$D_b F = [\phi, (a+b)*ab, (a+b)*bb + b]$	B	$\delta(A,a) = C$	0
$D_{aa} F = [\phi, (a+b)*ab + b, (a+b)*bb]$	C	$\delta(A,b) = D$	0
$D_{ab} F = [\lambda, (a+b)*ab, (a+b)*bb + b]$	D	$\delta(B,a) = D$	0.5
$D_{ba} F = [\phi, (a+b)*ab + b, (a+b)*bb]$	C	$\delta(B,a) = C$	0
$D_{bb} F = [\phi, (a+b)*ab, (a+b)*bb + b + \lambda]$	E	$\delta(B,b) = E$	0.2
$D_{aaa} F = [\phi, (a+b)*ab + b, (a+b)*bb]$	C	$\delta(C,a) = C$	0
$D_{aba} F = [\phi, (a+b)*ab + b, (a+b)*bb]$	C	$\delta(D,a) = C$	0
$D_{bba} F = [\phi, (a+b)*ab + b, (a+b)*bb]$	C	$\delta(E,a) = C$	0
$D_{aab} F = [\phi, (a+b)*ab + \lambda, (a+b)*bb + b]$	F	$\delta(C,b) = F$	0.3
$D_{abb} F = [\phi, (a+b)*ab, (a+b)*bb + b + \lambda]$	E	$\delta(D,b) = E$	0.2
$D_{bbb} F = [\phi, (a+b)*ab, (a+b)*bb + b + \lambda]$	E	$\delta(E,b) = E$	0.2
$D_{aaba} F = [\phi, (a+b)*ab + b, (a+b)*bb]$	C	$\delta(F,a) = C$	0
$D_{aabb} F = [\phi, (a+b)*ab, (a+b)*bb + b + \lambda]$	E	$\delta(F,b) = E$	0.2

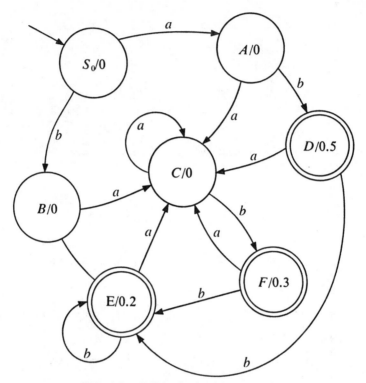

FIG. 4.3. DFA obtained from RFG_1.

and

   $\mu_{RFL}(ababb) = \min\ \left[0.3, 0.3, 0.3, 0.2, 0.5\right] = 0.2.$

Using RFG_2,

   $S_0 \to aA \to abD \to abaC \to ababF \xrightarrow{0.3} ababb.$

Example 4.4.2. Consider string ab. Using RFG_1,

   $S_0 \xrightarrow{0.5} aA \xrightarrow{0.5} ab$ and $S_0 \xrightarrow{0.7} aB \xrightarrow{0.4} ab$

so that $\mu_{RFL}(ab) = \max\ \left[\min(0.5, 0.5),\ \min(0.7, 0.4)\right] = 0.5$

Using RFG_2,

   $S_0 \to aA \xrightarrow{0.5} ab.$

4.5. RELATIONSHIP BETWEEN FORMAL LANGUAGES AND λ-FUZZY LANGUAGES

From the definition of the fuzzy language it is seen that if the elements of $F(\sigma)$, $\sigma \in \Sigma$, take only values 1 and 0, then the fuzzy automaton is a non-deterministic finite automaton. If in each row of matrix $F(\sigma)$ there is one 1 and the rest of the elements are all equal to 0, then the fuzzy automaton is a deterministic finite automaton. Therefore, finite-state deterministic and non-deterministic languages are special cases of λ-fuzzy language. The relationship between the formal languages and λ-fuzzy languages can be illustrated in Fig. 4.4 which is similar to the relationship between the formal languages and λ-probabilistic languages[12].

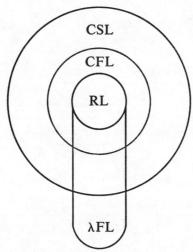

FIG. 4.4. Relationship between formal languages and fuzzy languages.

FL: λ-fuzzy languages

Fig. 4.4 shows that

RL ⊂ λFL
CFL ∩ λFL ≠ ∅
CSL ∩ λFL ≠ ∅

where RL, CFL, CSL, and λFL denote regular, context-free, context-sensitive, and λ-fuzzy languages, respectively, and $\emptyset$ denotes the empty set. The first relation is obvious. The second and third relations can be justified using the following example.

Example 4.5.1. Let $M = (\Sigma, S, F, \pi_0, \eta^{\lambda})$ where

$$\Sigma = \{1, 0\}$$

$$S = \{s_1, s_2\}$$

$$F(0) = \begin{bmatrix} 0.4 & 0.2 \\ 0.6 & 0.8 \end{bmatrix} \quad , \quad F(1) = \begin{bmatrix} 0.5 & 0.3 \\ 0.7 & 0.9 \end{bmatrix}$$

$$\pi_0 = \begin{bmatrix} 0.3 & 0.4 \end{bmatrix}$$

$$\eta^{\lambda} = \begin{bmatrix} 0.2 & 0.6 \end{bmatrix}$$

Consider a non-finite state, but context-free language

$$\{0^k \, 1^k, \, k = 1, 2, \ldots \}.$$

It can be shown that for any 2×2 matrix

$$A = \begin{bmatrix} a & b \\ c & d \end{bmatrix}$$

if $b < a < c < d$, define $A^n = \underbrace{A \circ A \circ \ldots \circ A}_{n \; A's}$

$$A^n = A$$

Therefore, for $k \geq 1$

$$F(0^k \, 1^k) = F(0^k) \circ F(1^k) = \big[F(0) \big]^k \circ \big[F(1) \big]^k = F(0) \circ F(1).$$

Evaluation of $F(0) \circ F(1)$

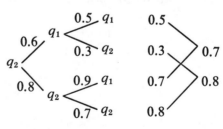

$$F(0^k \, 1^k) = F(0) \circ F(1) = \begin{bmatrix} 0.4 & 0.3 \\ 0.7 & 0.8 \end{bmatrix}$$

$$f_m(0^k\ 1^k) = \pi_0 \circ F(0^k\ 1^k) \circ \eta^{\mathcal{A}} = \begin{bmatrix} 0.28 & 0.32 \end{bmatrix} \circ \begin{bmatrix} 0.2 \\ 0.6 \end{bmatrix} = 0.192.$$

Therefore, $\{0^k\ 1^k, k = 1, 2, \ldots\}$ is acceptable by M with a cut-point $\lambda < 0.192$. This justifies that the intersection of CFL and λFL is non-empty.

Now consider a non-finite-state but context-sensitive language

$$\{0^k\ 1^k\ 0^k, k = 1, 2, \ldots\}$$

is applied to the same fuzzy automaton M. First we want to find $F(0^k\ 1^k\ 0^k)$. For the same reasoning mentioned above,

$$F(0^k\ 1^k\ 0^k) = F(0) \circ F(1) \circ F(0) = \begin{bmatrix} 0.4 & 0.3 \\ 0.7 & 0.8 \end{bmatrix} \circ F(0)$$

which may be found from the following diagram

Thus, for $k \geq 1$

$$F(0^k\ 1^k\ 0^k) = \begin{bmatrix} 0.4 & 0.3 \\ 0.6 & 0.8 \end{bmatrix}$$

and the grade of membership of $\{0^k\ 1^k\ 0^k, k = 1, 2, \ldots\}$ is

$$f_m(0^k\ 1^k\ 0^k) = \pi_0 \circ F(0^k\ 1^k\ 0^k) \circ \eta^{\mathcal{A}}$$

$$= \begin{bmatrix} 0.3 & 0.4 \end{bmatrix} \circ \begin{bmatrix} 0.4 & 0.3 \\ 0.6 & 0.8 \end{bmatrix} \circ \begin{bmatrix} 0.2 \\ 0.6 \end{bmatrix}$$

$$= \begin{bmatrix} 0.24 & 0.32 \end{bmatrix} \circ \begin{bmatrix} 0.2 \\ 0.6 \end{bmatrix} = 0.192 .$$

It is interesting to notice that

$$\{0^k \ 1^k \ 0^k, k = 1, 2, \dots \}$$

is acceptable by M with a cut-point $\lambda < 0.192$. This justifies the assumption that the intersection of CSL and λFL is non-empty.

REFERENCES

1. Zadeh, L. A. Fuzzy sets and systems. *Proc. Symp. on System Theory.* pp. 29–37. Polytechnic Press of the Institute of Brooklyn, New York, 1965.
2. Zadeh, L. A. Toward fuzziness in computer systems. *Fuzzy Algorithms and Languages.* University of California, Berkeley, Calif., 1969.
3. Mizumoto, M., Toyoda, J., and Tanaka, K. Fuzzy languages. *Trans. Elec. Commun. Eng. Japan* **53-C,** 333–40, 1970 (in Japanese).
4. Santos, E. S. Maximin automata. *Inform. Control.* 363–77, 1968.
5. Santos, E. S., and Wee, N. G. General formulation of sequential machines. *Inform. Control.* 5–10, 1970.
6. Lee, E. T., and Zadeh, L. A. Note on the fuzzy languages. *Inform. Sci.* 421–34, 1969.
7. Lee, E. T., and Zadeh, L. A. Fuzzy languages and their acceptance by automata. *Proc. 4th Princeton Conf. on Information Science and Systems,* p. 399, 1970.
8. Allen, A. D. Measuring the empirical properties of sets. *IEEE Trans. Syst. Man Cybernet.* **SMC-4,** 66–73, 1974.
9. Adams, E. W. Elements of a theory of unexact measurement. *Phil. Sci.* 205–28, 1968.
10. Adey, W. E. Organization of brain tissue: Is the brain a noisy processor? *Int. J. Neurosci.* 271–84, 1972.
11. Adaric, P. N., Borisov, A. N., Golender, V. E. An adaptive algorithm for recognition of fuzzy patterns. *Kibernet. Diagnost.* 13–18, 1968 (in Russian).
12. Boicescu, V. Sur les algebres de Lukasiewicz. *Logique, Automatique, Informatique,* pp. 71–97. Academie de la Republica Socialiste Roumanie, 1971.

CHAPTER 5

Applications of Fuzzy Concepts

5.1. INTRODUCTION

In the preceding chapters we developed the concepts of fuzzy logic and fuzzy functions by means of a suitable fuzzy algebra. The intention of this chapter is to discuss various applications of these concepts. In our quest for precision we show that imprecise concepts and fuzzy phenomena in the real world can be analysed and investigated via fuzzy models instead of fitting imprecise structures to mathematical models that make no provision for fuzziness.

Section 5.2 represents the subject of fuzzy neural networks, their analysis, relations to fuzzy automata, and their relations to fuzzy languages. Section 5.3 presents an approximation method via fuzzy functions which was originally investigated by C. L. Chang. Section 5.4 is represented, in part, as an argument supporting the representation of inexact concepts by fuzzy functions. In that section the transient behavior of binary switching circuits is studied under the assumption that the non-ideal transient behavior of switching components is best represented as a fuzzy phenomenon. In § 5.5 we introduce two examples of application of fuzzy matrices. The first one is in the area of pattern recognition and classification. It is quite clear to researchers in the field that most structures in pattern analysis are deliberately left fuzzy; thus the technique presented in this chapter illustrates a use of subjective evaluations as applied to a set of contemporary deductive problem-solving paradigms that deal with properties of elements in a pattern space. The other application of fuzzy matrices is illustrated in the area of role theory investigated by Thomason and Marinos.

The goal of this chapter is to illustrate some applications in the field and to invite the reader to open his field of study and let fuzzy sets penetrate through and give all of us the chance to explore new applications of fuzzy set theory in science and engineering. The interested reader will find that the literature cited in the bibliography contains many more applications of fuzzy sets in a variety of fields.

5.2. FUZZY NEURAL NETWORKS

In the nearly three decades since its publication the pioneering work of McCulloch and Pitts[1] has had a profound influence on the development of the theory of neural nets, in addition to stimulating much of the early work in automata theory and regular events[2-12].

Although the McCulloch–Pitts model of a neuron has contributed a great deal to the understanding of the behavior of neural-like systems, it fails to reflect the fact that the behavior of even the simplest type of nerve cell exhibits not only randomness but, more importantly, a type of imprecision which is associated with the lack of sharp transition from the occurrence of an event to its non-occurrence. It is possible that a better model for the behavior of a nerve cell may be provided by what might be called a *fuzzy* neuron, which is a generalization of the McCulloch–Pitts model. The concept of a fuzzy neuron employs some of the concepts and techniques of the theory of fuzzy sets which was introduced by Zadeh[13,14] and applied to the theory of automata by Wee and Fu[15], Tanaka *et al.*[16], Santos[17], and others. In effect, the introduction of fuzziness into the model of a neuron makes it better adapted to the study of the behavior of systems which are imprecisely defined by virtue of their high degree of complexity. Many of the biological systems, economic systems, urban systems, and, more generally, large-scale systems fall into this category.

In what follows, we shall present a preliminary account of a theory of fuzzy neural networks stressing its relations to automata and languages. Biologically oriented applications of this theory will be presented in subsequent papers.

5.2.1. Fuzzy Neurons

First, let us review the assumptions underlying the McCulloch–Pitts model of a cell (neuron)[1]. They are as follows.

(1) The activity of the cell is an "all-or-none" process.

(2) A certain fixed number of synapses must be excited within the period of latent addition in order to excite a cell at any time, and this number is independent of previous activity and position on the cell.

(3) The only significant delay within the cell is synaptic delay.

(4) The activity of any inhibitory synapse absolutely prevents excitation of the cell at that time.

(5) The structure of a neural network does not change with time.

Our model differs from the McCulloch–Pitts model mainly in the replacement of the first of the above assumptions with the following less restrictive assumption.

(1′) The activity of the cell is a "fuzzy" process.[1]

This leads us to the following definition.

Definition 5.2.1. A neuron is a *fuzzy neuron* if the following conditions are satisfied.

$$(1) \quad 0 \leqslant e_j(k), \qquad i_j(k) \leqslant 1 \tag{5.2.1}$$

where $e_j(k)$ stands for the jth excitatory input at time k, where the excitatory input is denoted by an arrow $\rightarrow$, $i_j(k)$ stands for the jth inhibitory input at time k, where the inhibitory input is denoted by a circle $-\circ$.

(2) The threshold of the neuron is a positive *real* number.

(3) The firing rules of the neurons are (i) all the inhibitory inputs must be 0, and (ii) the sum of all e_i's must be equal to or greater than the threshold T of the neuron:

$$\sum_i e_i^{(k)} \geqslant T. \tag{5.2.2}$$

When both of the above two firing rules are satisfied at time $t = k$ the neuron will fire at time $t = k + 1$; otherwise the fuzzy neuron remains in the quiet state.

(4) The outputs z_i of the neuron are equal to zero if it is quiet and equal to some positive real numbers μ_i, where

$$0 < \mu_i \leqslant 1 \tag{5.2.3}$$

if it is firing.

A fuzzy neuron is depicted in Fig. 5.1.

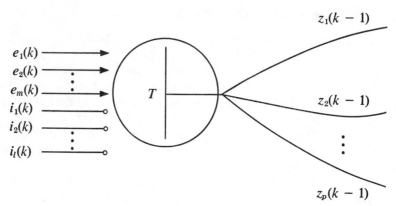

FIG. 5.1. A fuzzy neuron.

In what follows, we shall present a general synthesis procedure for realizing any fuzzy automaton using fuzzy neurons. As a preliminary we shall consider the analysis of networks composed of fuzzy neurons, namely the fuzzy neural networks.

5.2.2. Analysis of Fuzzy Neural Networks

It has been shown (18) that any (non-fuzzy) neural network is a finite-state automaton. The state of a neural network is determined by the Cartesian product of the states (firing or quiet) of the neurons. The derivation of a state diagram from a fuzzy neural network is best illustrated by the use of an example.

Consider the network of Fig. 5.3a. Assume that the initial state is the one where neuron A is firing (F) and neurons B and C are quiet (Q). In order to obtain the state-transition diagram, we must first find the *fuzzy state-transition tree* of the network which is described as follows.

(1) Apply the input symbol 0 to the initial state ($A = F$, $B = Q$, $C = Q$) and find the next states of the neurons A, B, and C. For example,

$$\left.\begin{array}{l} A = F \\ B = Q \\ C = Q \end{array}\right\} \quad \underrightarrow{\text{input} = 0} \quad \left\{\begin{array}{l} A = Q \\ B = Q \\ C = F \end{array}\right.$$

The next states of the neurons A, B, and C with an input 1 applied are obtained similarly. Display them as shown in Fig. 5.3b.

(2) Repeat (1). A branch b of level k will terminate if the k-level leaf associated with the branch b has appeared in the first k levels.

(3) The process ends when all the branches are terminated.

(4) The final step is to determine the grade of membership ρ of the state transition from the μ_i's, which is defined as follows.

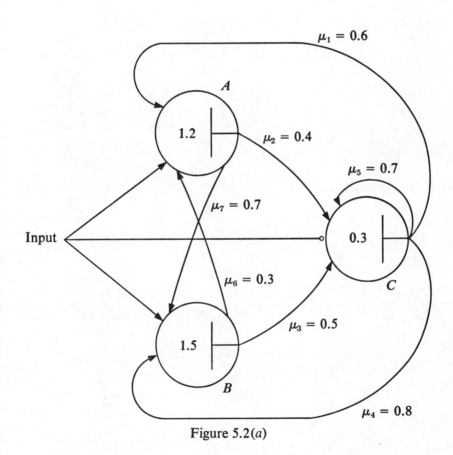

Figure 5.2(a)

Figure 5.2(*b*)

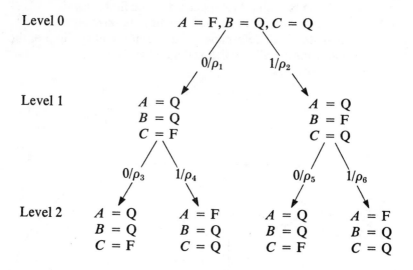

Level 0 $A = F, B = Q, C = Q$

$0/\rho_1$ $1/\rho_2$

Level 1 $A = Q$ $A = Q$
                 $B = Q$ $B = F$
                 $C = F$ $C = Q$

$0/\rho_3$ $1/\rho_4$ $0/\rho_5$ $1/\rho_6$

Level 2 $A = Q$ $A = F$ $A = Q$ $A = F$
           $B = Q$ $B = Q$ $B = Q$ $B = Q$
           $C = F$ $C = Q$ $C = F$ $C = Q$

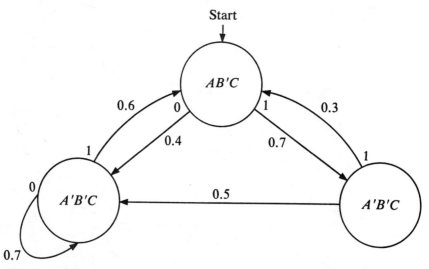

FIG. 5.2. (*a*) A neural network. (*b*) The state transition tree of the network. (*c*) The state diagram of the network.

Definition 5.2.2. Define the σ_i of a firing neuron as

$$\sigma_i = \text{Max} \left[\mu_1, \mu_2, \ldots, \mu_s \right] \tag{5.2.4}$$

and the σ_i of a quiet neuron to be zero, where μ_i are the inputs to the neuron at time $t = k$, excluding the input of the network at time $t = k$.

We postulate that the membership ρ of the state transition from the state at time k to the state at time $k + 1$ is given by

$$\rho = \text{Min} \left[\sigma_1, \ldots, \sigma_{m_f} \right] \tag{5.2.5}$$

where m_f is the number of firing neurons at time $t = k + 1$. The ρ so defined implies that each firing neuron has at least one input whose value is greater than or equal to ρ.

The complete fuzzy state-transition tree of the neural network of Fig. 5.2a is shown in Fig 5.2b and the grades of membership ρ_i of the state transitions are tabulated in Table 5.1. For simplicity we shall abbreviate a firing neuron and a quiet neuron by a capital letter with a prime. For instance the initial state is abbreviated by $AB'C'$. Using the states, the state-transition diagram of the network is given in Fig. 5.2c.

Table 5.1. The values of ρ_i of the state transition of the neural network of Fig. 5.2a.

ρ_1	0.4	ρ_4	0.6
ρ_2	0.7	ρ_5	0.5
ρ_3	0.7	ρ_6	0.3

It should be remarked here that the state-transition tree of a neural network containing n neurons can have *at most* 2^n levels, which in this example is $2^3 = 8$.

5.2.3. Synthesis of Fuzzy Automata by Fuzzy Neurons

A fuzzy automaton is based on a fuzzy process which is formally defined as follows.

Definition 5.2.3. A (finite) process in discrete time with a discrete state space $K = (q_1, q_2, \ldots, q_n)$ is called a (finite) *fuzzy process* if it satisfies the following conditions.

(1) The matrix $\mathbf{F}$ which describes the state transition has the following form:

$$\mathbf{F} = \begin{array}{c} \\ q_1 \\ q_2 \\ \vdots \\ q_n \end{array} \begin{array}{cccc} q_1 & q_2 & \cdots & q_n \\ \left[\begin{array}{cccc} \rho_{11} & \rho_{12} & \cdots & \rho_{1n} \\ \rho_{21} & \rho_{22} & \cdots & \rho_{2n} \\ \vdots & \vdots & & \vdots \\ \rho_{n1} & \rho_{n2} & \cdots & \rho_{nn} \end{array}\right] \end{array}$$

where $0 \leqslant \rho_{ij} \leqslant 1$ denotes the grade of membership of state transition from state q_i to state q_j. This matrix will be called the *fuzzy state-transition matrix* of the fuzzy process.

(2) Let M be a fuzzy set defined on K and

$$w_M^{(0)} = \left[\eta_{q_1}^{(0)}, \eta_{q_2}^{(0)}, \ldots, \eta_{q_n}^{(0)}\right]$$

be a row vector called the *initial state designator of A*, where $\eta_{q_1}^{(0)}$ is the grade of membership of q_i with respect to M at time $t = 0$. Then the state designator of M at $t = k$,

$$w_M^{(k)} = \left[\eta_{q_1}^{(k)}, \eta_{q_2}^{(k)}, \ldots, \eta_{q_n}^{(k)}\right]$$

is obtained by

$$w_M^{(k)} = w_M^{(0)} \circ \underbrace{(\mathbf{F} \circ \mathbf{F} \circ \ldots \circ \mathbf{F})}_{(k-1) \text{ operations}} = w_M^{(0)} \circ \mathbf{F}^K = w_M^{(0)} \circ \left[\rho_{ij}^{(k)}\right] \qquad (5.2.6)$$

where

$$\rho_{ij}^{(k)} = \underset{\substack{\text{over all parallel} \\ \text{paths from } q_i \text{ to } q_j \\ \text{with } k-1 \text{ numbers} \\ \text{of transitions}}}{\max} \left\{ \underset{\substack{\text{over all series} \\ \text{paths from } q_i \text{ to } q_j \\ \text{with } k-1 \text{ numbers} \\ \text{of transitions}}}{\min} \left\{ \rho_{il_i}, \rho_{l_1 l_2}, \ldots, \rho_{l_{(k-1)}j} \right\} \right\}.$$

The following remarks are in order.

(1) The fuzzy state-transition matrix $\mathbf{F}$ of a fuzzy process resembles somewhat the stochastic matrix of a Markov chain for $0 \leqslant \rho_{ij} \leqslant 1$; however, in general $\sum_{k=1}^{n} \rho_{ik} = 1$, for all i.

(2) The reason we use the max min [equation (5.2.7)] rule in defining $\rho_{ij}^{(k)}$ is that the state transition from q_i to q_j in a fuzzy process may be considered as water (gas, electricity, traffic flow, etc.) flow through a water supply system of which the water pipes are series–parallel interconnected.

(3) When ρ_{ij} takes only two values 0 and 1, the process becomes a *non-deterministic process*. In addition, if only any one element of each

row of matrix $\mathbf{F}$ is 1 and the remaining elements of each row are equal to 0, then the process is a *deterministic process*.

Let Σ^* denote the set of all input sequences (tapes) including the empty word ϵ over a finite set of alphabets Σ. Then the domain of $\mathbf{F}$ is extended from Σ to Σ^* by defining $\mathbf{F}(\epsilon) = I_n$, where ϵ denotes the empty word and I_n is an $n \times n$ identity matrix:

$$\mathbf{F}(\sigma_{i_1}\sigma_{i_2} \ldots \sigma_{i_k}) = \mathbf{F}(\sigma_{i_1}) \circ \mathbf{F}(\sigma_{i_2}) \circ \ldots \circ \mathbf{F}(\sigma_{i_k})$$

$$= \left[\rho_{ij}^{(k)} \right]$$

$$k \geqslant 2 \quad \text{and} \quad \sigma_{i_j} \in \Sigma, j = 1, 2, \ldots, k. \tag{5.2.8}$$

The symbol $\circ$ denotes the max min operation, and $\rho_{ij}^{(k)}$ is defined as in equation (5.2.7).

In the previous section, it has been shown that a fuzzy neural network is a fuzzy finite state automaton (FFSA) or finite fuzzy automaton (FFA). Then it is natural to ask the question: For a given fuzzy automaton transition diagram, can we always find a fuzzy neural network realizing it? The answer is yes. We shall begin this discussion by introducing the following definitions.

Definition 5.2.4. Let M be an FFA and q be a state of M. State q is said to be *homogeneous* if all the state-transition lines of the state diagram of M incident to q are driven by the same input; otherwise we call it a *non-homogeneous state*. If the input space is $(0,1)$, the homogeneous states driven by 1 and 0 are called the *1-state* and *0-state*, respectively. For example, in Fig. 5.2c the states $AB'C'$ and $A'BC'$ are 1-states and the states $A'B'C$ is a 0-state.

Definition 5.2.5. An FFA M is *homogeneous* if all its states are homogeneous; otherwise it is *non-homogeneous*.

For example, the automaton of Fig. 5.2c is a homogeneous automaton.

Synthesis of state transition diagram

Part A–I. Synthesis of homogeneous FFA. The synthesis procedure is illustrated by the following example.

Example 5.2.1. Consider the automaton M_1, described by the state diagram of Fig. 5.3a.

(1) For a homogeneous automaton M with n states $A, B, C, \ldots$, use n neurons. Each of the neurons is labeled by $A, B, C, \ldots$.

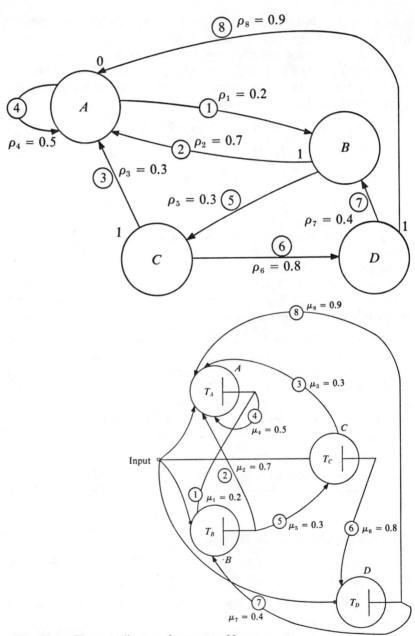

FIG. 5.3. (a) The state diagram of automaton M_1.
(b) Its fuzzy neural network realization.

(2) Classify the 1-states and 0-states of the automaton: A, 1-state; B, C, D, 0-states.

(3) Define the identity mapping ϕ from the set of n states onto the set of n neurons as $\phi(X) = X$, $X = A, B, C, \ldots$.

(4) Start with the input node, and draw a line to $\phi(X)$ with an excitatory input if X is a 1-state and with an inhibitory input if X is a 0-state. We shall call the neuron $\phi(X)$ a 1-neuron or a 0-neuron if X is a 1-state or a 0-state, respectively.

(5) Draw additional lines with excitatory inputs from the output of $\phi(X)$ to the input of $\phi(Y)$ for all state transitions from state X to state Y.

(6) Define a one-to-one mapping ϕ from the n states of the automaton into the states of the n neurons as follows.

The n states of the automaton		The states of the n neurons
A	$\xrightarrow{\phi}$	$A\ B'C' \ldots$
B	$\xrightarrow{\phi}$	$A'B\ C' \ldots$
C	$\xrightarrow{\phi}$	$A'B'C\ \ldots$

(7) Determine the threshold values of the neurons. The way of determining the threshold values of the neurons is to construct the state-transition table of the neural network of Fig. 5.3b according to the state-transition diagram of M_1.

If state X to state Y is by an input a (either 0 or 1), then the state of the n neurons $\phi(X)$ to $\phi(Y)$ is by the same input a. This is shown in Table 5.2. We find that the ranges of the threshold values of the fuzzy neurons are

$$0.9 < T_A \leq 1.3$$

$$0 < T_B \leq 0.2$$

$$0 < T_C \leq 0.3$$

$$0 < T_D \leq 0.8 \ .$$

The fuzzy neural network realization of automaton M_1 is shown in Fig. 5.3b.

The way of determining the threshold values of the neurons described in step (7) of the above procedure is rather tedious. We shall prove that it is unnecessary.

Table 5.2. The state-transition table and the requirement on the threshold values of the neurons of the neural network of Fig. 5.3*b*

Present state	Input 0		Input 1	
	Next state	Requirement on the value T of the threshold	Next state	Requirement on the value T of the threshold
A	A'	$T_A > 0.5$	A	$T_A \leq 1.5$
B'	B	$T_B \leq 0.2$	B'	No information[1]
C'	C'	$T_C > 0$	C'	No information
D'	D'	$T_D > 0$	D'	No information
A'	A'	$T_A > 0.7$	A	$T_A \leq 1.7$
B	B'	$T_B > 0$	B'	No information
C'	C	$T_C \leq 0.3$	C'	No information
D'	D'	$T_D > 0$	D'	No information
A'	A'	$T_A > 0.3$	A	$T_A \leq 1.3$
B'	B'	$T_B > 0$	B'	No information
C	C'	$T_C > 0$	C'	No information
D'	D	$T_D \leq 0.8$	D'	No information
A'	A'	$T_A > 0.9$	A	$T_A \leq 1.9$
B'	B	$T_B \leq 0.4$	B'	No information
C'	C'	$T_C > 0$	C'	No information
D	D'	$T_D > 0$	D'	No information

1. No information about the requirement of the value of the neuron is provided when it has an inhibitory input with input value 1.

Theorem 5.2.1. In the realization of a homogeneous automaton the range of the threshold values of a 1-neuron is

$$\mu_{max} < T_{\text{1-neuron}} \leq 1 + \mu_{min} \qquad (5.2.9)$$

when there is only one 1-neuron in the neural network and

$$1 < T_{\text{1-neuron}} \leq 1 + \mu_{min} \qquad (5.2.10)$$

when there is more than one 1-neuron in the neural network, and the range of the threshold values of a 0-neuron is

$$0 < T_{0\text{-neuron}} \leq \mu_{min} \tag{5.2.11}$$

where

$$\mu_{min} = \min_{\substack{\text{all excitatory} \\ \text{inputs } \mu_i \text{ to the} \\ \text{neuron}}} [\mu_1, \mu_2, \ldots, \mu_v] \tag{5.2.12}$$

$$\mu_{max} = \max_{\substack{\text{all excitatory} \\ \text{inputs } \mu_i \text{ to the} \\ \text{neuron except to} \\ \text{the neural network}}} [\mu_1, \mu_2, \ldots, \mu_r]. \tag{5.2.13}$$

Proof. We shall prove that the values of $T_{1\text{-neuron}}$ and $T_{0\text{-neuron}}$ cannot be outside the ranges described above. First consider the 0-neuron case shown in Fig. 5.4a. When the input is 1, i.e. the inhibitory input having input value 1, the neuron never fires. In this case no information is provided in determining the threshold value of the neuron. Consider the case where the input of the neural network is 0. There is at most one input to the 0-neuron with input value other than 0, because at each moment one and only one neuron of the neural network can be in the firing state. The threshold value of the 0-neuron cannot be 0, because otherwise it fires all the time. This implies that the automaton always remains in the same state, since at each moment one and only one neuron can be in the firing state. An automaton corresponding to a neural network containing such a neuron is a trivial automaton (one-state automaton) from that moment on. So we eliminate the value 0 as the threshold. The threshold value of the 0-neuron cannot be greater than μ_{min} either, since otherwise there will exist some excitatory input to the 0-neuron for which the 0-neuron will not fire. This will make the fuzzy neural network unable to realize the given automaton. When the threshold value is in the range $0 < T_{0\text{-neuron}} \leq \mu_{min}$ the neuron can provide both quiet and firing states in accordance with the transition diagram of the automaton.

Now consider the 1-neuron as shown in Fig. 5.4b,c. The reason that the threshold value cannot be greater than $1 + \mu_{min}$ is similar to the case just discussed in which the threshold value of the 0-neuron cannot be greater than μ_{min}.

When the input is 0 for any 1-neuron the next state must be quiet, because the next state of the automaton cannot be in the state corre-

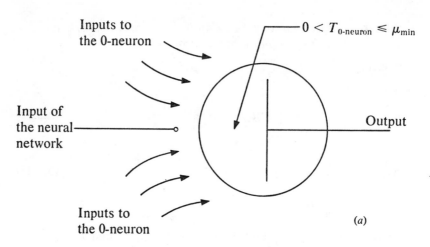

(a)

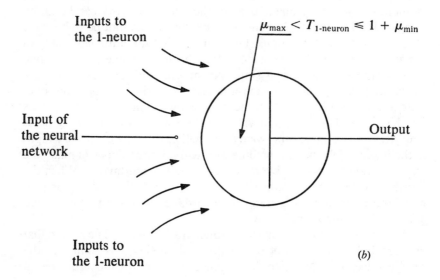

(b)

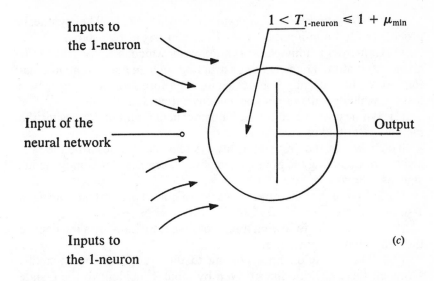

FIG. 5.4. (*a*) The range of the threshold value of a 0-neuron. (*b*) The range of the threshold value of a 1-neuron when there is only one 1-neuron in the neural network. (*c*) The range of the threshold value of a 1-neuron when there is more than one 1-neuron in the neural network.

sponding to this 1-neuron. For this reason the threshold cannot be less than μ_{max}. When there is *only one* 1-neuron in the network and its threshold value is in the range $\mu_{max} \leqslant T_{1\text{-neuron}} \leqslant 1 + \mu_{min}$, it can provide both quiet and firing states depending upon the state transition of the automaton. However, when there is *more than one* 1-neuron, the range of the threshold values of the 1-neuron is then $1 < T_{1\text{-neuron}} \leqslant 1 + \mu_{min}$. Therefore, step (7) of the synthesis procedure can be replaced by step (7′).

(7′) Determine the threshold values of the neurons. The range of the threshold values of a 1-neuron is $\mu_{max} < T_{1\text{-neuron}} \leqslant 1 + \mu_{min}$ when there is only one 1-neuron in the neural network and

$$1 < T_{1\text{-neuron}} \leqslant 1 + \mu_{min}$$

when there is more than one 1-neuron in the neural network and the range of the threshold values of a 0-neuron is $0 < T_{0\text{-neuron}} \leqslant \mu_{min}$. For example, in the neural network of Fig. 5.3*b* the A neuron is a 1-neuron with thresholds T_A, and the B, C, and D neurons are 0-neurons with threshold T_B, T_C, and T_D respectively, which agrees with the theorem.

Part A-II. Synthesis of non-homogeneous FFA. The synthesis procedure for a non-homogeneous FFA consists of two steps.

(1) Construct a homogeneous automaton which is equivalent to the given automaton. The way of constructing such an equivalent automaton is to split each non-homogeneous state into two, a 1-state and a 0-state, which results in a homogeneous automaton.

(2) Synthesize the equivalent homogeneous automaton using the method described in Part A-I.

To show that the first step is always feasible, we describe the procedure for obtaining the homogeneous states from a non-homogeneous state as follows.

(1) Split the non-homogeneous state into two states: one is a 1-state, and the other is a 0-state.

(2) Consider the following three possible transition lines incident to the non-homogeneous state.

(2.1) The transition lines coming to the state from other states. Draw all the transition lines driven by input 1 incident to the 0-state with the values of ρ_i's unchanged.

(2.2) The transition lines going out of the state. For each transition line going out of the state driven by an input which is either 0 or 1 draw two lines driven by the same input to that state with the same membership ρ_i if the state to which the transition line is incident is a homogeneous state. If the state to which the transition line is incident is a non-homogeneous state, draw two lines driven by the same input with the same membership ρ_i to the equivalent 1-state or 0-state of that state depending on whether the input is 1 or 0.

(2.3) The self-looped transition lines of the state. If a self-looped transition line is driven by 1, draw a self-looped transition line driven by 1 at the 1-state and a transition line from the 0-state driven by 1 to the 1-state. If a self-looped transition line is driven by 0, draw a self-looped transition line driven by 0 at the 0-state and a transition line from the 1-state driven by 0 to the 0-state. Again, the memberships ρ_i of all the state transitions are kept the same.

This procedure is illustrated by the following example.

Example 5.2.2. An automaton M_2 is described by the state transition diagram of Fig. 5.5a in which states A, B, and D are non-homogeneous states and state C is a homogeneous 0-state. States A, B, and D are split into two states A_1 and A_0, B_1 and B_0, D_1 and D_0, respectively. The subscripts 1 and 0 denote the 1-state and the 0-state.

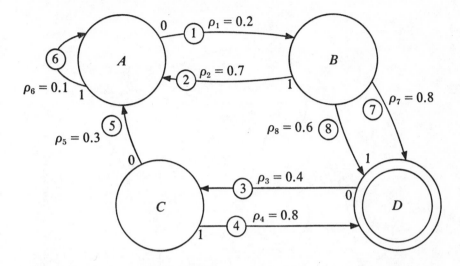

The state-transition diagram of the homogeneous automaton M_2' in Fig. 5.5b is constructed by the procedure just described.

Definition 5.2.6. Two states of a fuzzy automaton are *equivalent* if they are (1) equivalent in the ordinary sense, i.e. for all input sequences applied to the two states outputs are identical, and (2) the grades of acceptance to all input sequences starting from the two states are identical.

Definition 5.2.7. Two fuzzy automata M_a and M_b are *equivalent* if for every state of M_a there is at least one equivalent state in M_b and *vice versa*.

We can easily show that the following theorem holds.

Theorem 5.2.2. Let M be a non-homogeneous FFA.

(a) For any non-homogeneous state A of M, A_1 and A_0 are always obtainable.

(b) States A_1, A_0, and A are equivalent.

(c) If B is a homogeneous state of M, then B remains homogeneous under the transformation from the non-homogeneous state to the homogeneous state described above.

Example 5.2.2. (cont.) The equivalence relation among the states of the two automata: automaton M_2 can be best seen from the state-transition tables of the two automata which are shown in Table 5.3. In

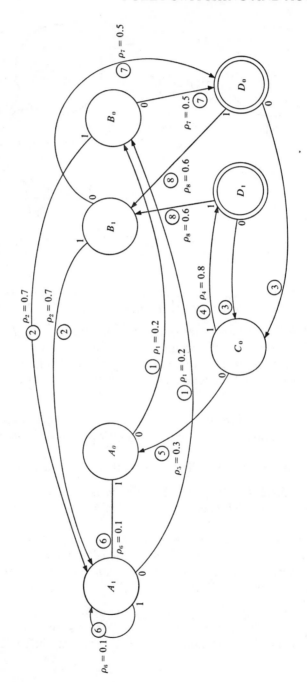

FIG. 5.5. (*a*) A non-homogeneous automaton M_2. (*b*) The equivalent homogeneous automaton M_2'.

this table, besides the next states, the memberships of the state transitions are also shown. For example the entity $A_1/0.1$ at the first row and the first column of the left-hand table means that the membership of the state transition from state A_1 to state A_1 to state A_1 by the input 1 is 0.1. It is seen from Table 5.3 that states A_1, A_0, and A, B_1, B_0, and B, and D_1, D_0, and D are equivalent states. Since for every state of automaton M_2 there is at least one equivalent state in the homogeneous automaton M_2' and *vice versa*, the two automata are equivalent.

Corollary 5.2.1. For any non-homogeneous automaton there exists a homogeneous automaton which is equivalent to it.

Example 5.2.2. (cont.) Realize the non-homogeneous automaton M_2 of Fig. 5.5a using a fuzzy neural network.

The equivalent homogeneous automaton of M_2 was obtained above. Applying the method described in Part A-I, the realization is shown in Fig. 5.5b in which the output of the neural network is taken from the outputs of neurons D_1 and D_0 feeding to the input of an OR neural network. The ranges of the threshold values of the fuzzy neurons are obtained by Theorem 5.2.1 and are shown in Table 5.4.

Table 5.3. The state transition tables of M_2 and its homogeneous automaton equivalent M_2'

q_k \ x_k	1	0	Equivalent class	q_k \ x_k	1	0	Equivalent class
A_1 A_0	$A_1/0.1$ $A_1/0.1$	$B_0/0.2$ $B_0/0.2$	I	A	$A/0.1$	$B/0.2$	I
B_1 B_0	$A_1/0.7$ $A_1/0.7$	$(D_0)/0.5$ $(D_0)/0.5$	II	B	$A/0.7$	$(D)/0.5$	II
C_0	$(D_1)/0.8$	$A_0/0.3$	III	C	$(D)/0.8$	$A/0.3$	III
(D_1) (D_0)	$B_1/0.6$ $B_1/0.6$	$C_0/0.4$ $C_0/0.4$	IV	(D)	$B/0.6$	$C/0.4$	IV

The following theorem concerns the lower bound of the number of neurons required for synthesizing an n-state *minimal* automaton, that is in this minimal automaton no two states are equivalent.

Theorem 5.2.3. For an *n-state minimal automaton, there must be at least* $\lceil \log_2 n \rceil$ [2] *number of neurons to realize it.*

Proof. Since a neural network containing M neurons can have at most 2^m states, 2^m must be greater than or equal to n. Since m must be an integer, $m \geq \lceil \log_2 n \rceil$.

Corollary 5.2.2. Let m denote the number of neurons required to realize an *n*-state minimal machine by the method described in this section. Then such a realization with $\lceil \log_2 n \rceil \leq m \leq 2n$ can always be obtained.

2. The symbol $\lceil \log_2 n \rceil$ denotes the least integer which is greater than or equal to $\log_2 n$.

Table 5.4. The ranges of threshold values of the neurons of the realization of M_2

Threshold	Lower bound: the threshold is greater than	Upper bound: the threshold is less than or equal to
T_{A_1}	1	1.1
T_{A_0}	0	0.3
T_{B_1}	1	1.6
T_{B_0}	0	0.2
T_{D_1}	1	1.8
T_{D_0}	0	0.5
T_{C_0}	0	0.4

Part B. Realization of the grade of acceptance of the FFA. The synthesis of homogeneous and non-homogeneous state transition diagrams using fuzzy neurons was presented in Part A. In this part we shall present the realization of the degree of acceptance of the FFSA obtained in Part A. The grade of acceptance is defined in Definition (5.2.7). The problem is to evaluate the grade of acceptance. The method for evaluating the grade of acceptance is best illustrated by the use of an example.

Example 5.2.3. Consider the FFA M_2 of Example 5.2.2:

$$M_2 = (K, \Sigma, \delta, q_0, F)$$

where

$$K = (A, B, C, D)$$
$$\Sigma = (0,1)$$
$$q_0 = \{(A,0.6), (B,0.7), (C,0.8), (D,0.9)\}$$
$$F = \{D\}$$

and δ is the fuzzy mapping from $K \times \Sigma$ to K as represented by $M(1)$ and $M(0)$ where

$$M(1) = \begin{array}{c} \\ A \\ B \\ C \\ D \end{array} \begin{array}{cccc} A & B & C & D \\ \left[\begin{array}{cccc} 0.1 & 0 & 0 & 0 \\ 0.7 & 0 & 0 & 0 \\ 0 & 0 & 0 & 0.8 \\ 0 & 0.6 & 0 & 0 \end{array}\right] \end{array}$$

and

$$M(0) = \begin{array}{c} \\ A \\ B \\ C \\ D \end{array} \begin{array}{cccc} A & B & C & D \\ \left[\begin{array}{cccc} 0 & 0.2 & 0 & 0 \\ 0 & 0 & 0 & 0 \\ 0.3 & 0 & 0 & 0 \\ 0 & 0 & 0.4 & 0 \end{array}\right] \end{array}$$

We have shown that M_2' of Fig. 5.5*b* is equivalent to M_2.

$$M_2' = (K', \Sigma, \delta', q_0', F')$$

where

$$\Sigma = (0,1) \cdot$$
$$K' = (A_1, A_0, B_1, B_0, C_0, D_1, D_0)$$
$$q_0' = \{(A_1,0.6), (A_0,0.6), (B_1,0.7), (B_0,0.7),$$
$$(C_0,0.8), (D_1,0.9), (D_0,0.9)\}$$
$$F' = (D_1, D_0)$$

and δ' is the fuzzy mapping from $K' \times \Sigma$ to K' as represented by $M'(1)$ and $M'(0)$, where

$$M'(1) = \begin{array}{c} \\ A_1 \\ A_0 \\ B_1 \\ B_0 \\ C_0 \\ D_1 \\ D_0 \end{array} \begin{array}{ccccccc} A_1 & A_0 & B_1 & B_0 & C_0 & D_1 & D_0 \\ \begin{bmatrix} 0.1 & 0 & 0 & 0 & 0 & 0 & 0 \\ 0.1 & 0 & 0 & 0 & 0 & 0 & 0 \\ 0.7 & 0 & 0 & 0 & 0 & 0 & 0 \\ 0.7 & 0 & 0 & 0 & 0 & 0 & 0 \\ 0 & 0 & 0 & 0 & 0 & 0.8 & 0 \\ 0 & 0 & 0.6 & 0 & 0 & 0 & 0 \\ 0 & 0 & 0.6 & 0 & 0 & 0 & 0 \end{bmatrix} \end{array}$$

$$M'(0) = \begin{array}{c} \\ A_1 \\ A_0 \\ B_1 \\ B_0 \\ C_0 \\ D_1 \\ D_0 \end{array} \begin{array}{ccccccc} A_1 & A_0 & B_1 & B_0 & C_0 & D_1 & D_0 \\ \begin{bmatrix} 0 & 0 & 0 & 0.2 & 0 & 0 & 0 \\ 0 & 0 & 0 & 0.2 & 0 & 0 & 0 \\ 0 & 0 & 0 & 0 & 0 & 0 & 0.5 \\ 0 & 0 & 0 & 0 & 0 & 0 & 0.5 \\ 0 & 0.3 & 0 & 0 & 0 & 0 & 0 \\ 0 & 0 & 0 & 0 & 0.4 & 0 & 0 \\ 0 & 0 & 0 & 0 & 0.4 & 0 & 0 \end{bmatrix} \end{array}$$

The procedure for evaluating the grade of acceptance is as follows: Suppose $X = 100$. The grade of acceptance for

$$q'_0 = (A_1, 0.6), (A_0, 0.6), (B_1, 0.7), (B_0, 0.7),$$
$$(C_0, 0.8), (D_1, 0.9), (D_0, 0.9)$$

$$x = 100$$

and

$$F' = (D_1, D_0)$$
$$\mu_1 = (0.6, 0.6, 0.7, 0.7, 0.8, 0.9, 0.9)$$

so

$$\mu_4 = \mu_1 M'(100)$$

$$\mu_4 = \left[\left[[\mu_1 M'(1)] M'(0) \right] M'(0) \right].$$

(1) At time $t = 1$, set the initial values in the registers $R_{A_1}, \ldots, R_{D_0}$ (see Fig. 5.6) to be the values of the initial designator, namely

$$\mu_1 = \left[r_{A_1}^{(1)}, r_{A_0}^{(1)}, r_{B_1}^{(1)}, r_{B_0}^{(1)}, r_{C_0}^{(1)}, r_{D_1}^{(1)}, r_{D_0}^{(1)} \right]$$
$$= \left[0.6, 0.6, 0.7, 0.7, 0.8, 0.9, 0.9 \right]$$

where $r_{A_1}^{(1)}$ is the value in R_{A_1} at time $t = 1$, and $r_{A_0}^{(1)}, \ldots, r_{D_0}^{(1)}$ are similarly defined, the superscript denoting the time. Note that the row vector $\left[r_{A_1}^{(k)}, r_{A_0}^{(k)}, r_{B_1}^{(k)}, r_{B_0}^{(k)}, r_{C_0}^{(k)}, r_{D_1}^{(k)}, r_{D_0}^{(k)}\right]$ denotes the grade of membership of the states at time k.

(2) At time $t = 2$, if the input is 1, set all the excitatory inputs to the 0-neurons to be zero. If the input is 0, set all the excitatory inputs to the 1-neurons to be zero. Define

$$(v_x)_{\min} = \min \left[\mu_x, r_x^{(1)}\right] \tag{5.2.13}$$

where μ_x comes from the register R_x with the value $r_x^{(1)}$. Replace the $r_x^{(1)}$ by the maximum of the $(v_x)_{\min}$'s which are incident to the neuron x. Denote this value by $r_x^{(2)}$.

(3) Repeat step 2 $(K - 1)$ times, where K is the length of the input sequence. The μ_k are thus obtained. For our example μ_2, μ_3, μ_4 are tabulated in Table 5.5.

It is of interest to note that if the input at time $t = k$ is 1, the values $r_{0\text{-neuron}}^{(k+1)}$ in all the registers of the 0-neurons are zero. If the input at time $t = k$ is 0, the values $r_{1\text{-neuron}}^{(k+1)}$ in all the registers of the 1-neurons are zero.

After the μ_k are obtained, the grade of acceptance can be obtained by using a neuron OR network as shown in Fig. 5.6. It should be remarked that one additional unit delay is introduced by the OR neural network. The complete realization of M_2 is shown in Fig. 5.6.

5.2.4. Fuzzy Language Recognizers

The neural realization of FFA presented in the previous section can be used as fuzzy language recognizers. This is demonstrated by the following example.

Table 5.5. The sequences of the values in the registers

$$\mu_k = \left[r_{A_1}^{(k)}, r_{A_0}^{(k)}, r_{B_1}^{(k)}, r_{B_0}^{(k)}, r_{C_0}^{(k)}, r_{D_1}^{(k)}, r_{D_0}^{(k)}\right]$$

where $r_{A_1}^{(k)}$ is the value in R_{A_1} at time k. The $r_{A_0}^{(k)}, \ldots, r_{D_0}^{(k)}$ are similarly defined.

Time	R_{A_1}	R_{A_0}	R_{B_1}	R_{B_0}	R_{C_0}	R_{D_1}	R_{D_0}	μ_k
$t = 1$	0.6	0.6	0.7	0.7	0.8	0.9	0.9	μ_1
$t = 2$	0.7	0	0.6	0	0	0.8	0	μ_2
$t = 3$	0	0	0	0.2	0.4	0	0.5	μ_3
$t = 4$	0	0.3	0	0	0.4	0	0.2	μ_4

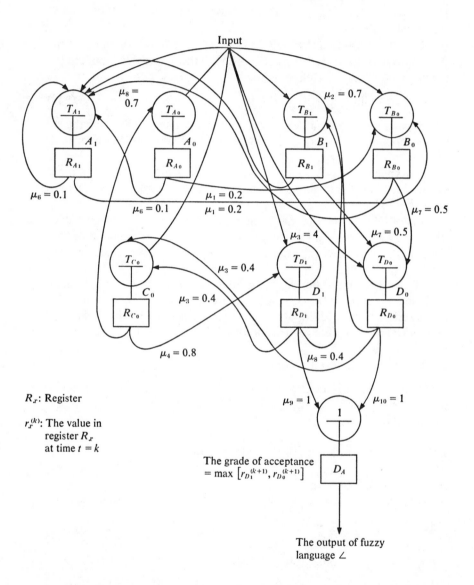

FIG. 5.6. The complete realization of M_2: R_x, register; $T_x^{(k)}$, the value in the register R_x at time $t = k$; grade of acceptance is given by

$$\max \left[r_{D_1}^{(k+1)}, r_{D_0}^{(k+1)} \right].$$

Example 5.2.4. Consider the FFA M_2. Let L be a fuzzy language generated by the FFA M_2. Suppose there are two sequences of input symbols $x = 100$ and $y = 1001$. What is the grade of acceptance of x and y in L?

It is seen from Table 5.5 that the grade of acceptance of x is 0.2:

$$F' = (0, 0, 0, 0, 0, 1, 1)$$

$$Q_4 = (0, 0.3, 0, 0, 0.4, 0, 0.2) \qquad \text{from Table 5.5}$$

$$\mu_M(x) = \mu_M(100)$$

$$= \max_{q_j} \mu_{F' \cap Q_4}(q_j)$$

$$= 0.2.$$

If one additional input symbol 1 is applied, then

	R_{A_1}	R_{A_0}	R_{B_1}	R_{B_0}	R_{C_0}	R_{D_1}	R_{D_0}
$t = 5$	0.1	0	0.2	0	0	0.4	0 .

Thus the grade of acceptance of $y = 1001$ is

$$\mu_M(y) = \mu_M(1001)$$

$$= \max_{q_j} \mu_{F' \cap Q_5}(q_j)$$

$$= 0.4.$$

It should be clear that the fuzzy neural network of Fig. 5.6 which realizes FFA M_2 is the fuzzy language recognizer for the fuzzy language accepted by M_2.

5.3. APPROXIMATION OF FUNCTIONS BY FUZZY FORMS

Since any real function on an n-dimensional Euclidean space can be approximated by a piecewise linear function with any desired accuracy, the concept of an *inexact function* provides a convenient tool for function approximation by noting that any piecewise linear function is really an inexact function. This is very important, because by this simple representation piecewise linear or non-linear functions can be more thoroughly studied.

Here we shall consider only approximations of real functions on an n-dimensional Euclidean space E^n. Let $y = f(x)$ be the function to be approximated, where $y \in E^1$ and $x \in E^n$. The usual approach is to observe a finite number of pairs of values $(y^1, x^1), (y^2, x^2), \ldots, (y^N, x^N)$, where $y^i = f(x^i)$ for $i = 1, \ldots, N$, and to fit these pairs of values by some function $F(X, \theta)$ of known form, e.g. $F(x, \theta)$ may be a linear function. The fitting is usually done by the so-called least-squares procedure; that is, we define the error sum of squares

$$S(\theta) = \sum_1^N [y^i - F(x^i, \theta)]^2$$

and find a value θ^* of θ which minimizes $S(\theta)$. We then use $F(x, \theta^*)$ as an approximation of $f(x)$. The above study is often called regression analysis. When $F(x, \theta)$ is linear in the parameter θ, the study is called linear regression analysis. However, it is often found that data can not be fitted very well by linear functions. If one uses higher-order functions by like polynomials, then, in order to have a good fit, these polynomials may be very complicated for data of large dimensions. In order to circumvent these difficulties, we shall choose $F(x, \theta)$ to be an *inexact function*. In particular, we shall let $F(x, \theta)$ be an inexact function in terms of linear functions. In this case the study is called *inexact regression analysis* or piecewise linear regression analysis.

We now can formulate the inexact regression analysis as follows. Without loss of generality, we can assume that the dimension n of the Euclidean space E^n is greater than 1, and the nth component x_n of a vector $x = (x_1, \ldots, x_n)$ is always equal to 1. Let $\theta^i = (\theta^i_1, \ldots, \theta^i_n)$. Then, we define

$$h_i(x, \theta^i) = \theta^i \cdot x \qquad \text{for } i = 1, \ldots, m$$

where $h_1, \ldots, h_m$ are linear functions of x. The inexact regression analysis is to select an inexact function $F(x, \theta^1, \ldots, \theta^m)$ in terms of $h_1, \ldots, h_m$ such that

$$S(\theta^1, \ldots, \theta^m) = \sum_1^N [y^i - F(x^i, \theta^1, \ldots, \theta^m)]^2$$

is minimized.

There are many inexact function terms of $h_1, \ldots, h_m$. However, as discussed in the previous section, every inexact function is a fuzzy function in a conjunctive normal form. For example,

$$F(x, \theta^1, \ldots, \theta^m) = h_1(x, \theta^1) + h_2(x, \theta^2) + \ldots + h_m(x, \theta^m)$$

is such an inexact function. Note that $+$ is interpreted as Max, i.e. the fuzzy algebra will be used in our inexact regression analysis. In general, the procedure of inexact regression analysis can be stated as follows.

Step 1: Decide the number m of basic linear functions.

Step 2: Select an inexact function $F(x, \theta^1, \ldots, \theta^m)$ in conjunctive normal form.

Step 3: Find values $\theta^1_*, \ldots, \theta^m_*$ of $\theta^1, \ldots, \theta^m$ which minimize $S(\theta^1, \ldots, \theta^m)$.

After Step 3 is completed, we then use $F(x, \theta^1, \ldots, \theta^m)$ as an approximation of $f(x)$. Step 1 and Step 2 are related to model selection of regression analysis. They are usually done by experience. Step 3 is a minimization problem.

We shall now consider an algorithm for finding the minimum of the error square sum $S(\theta^1, \ldots, \theta^m)$ defined above. There exist many procedures for solving general unconstrained minimization problems. For example, the gradient method and Newton–Raphson method are the most frequently used iterative procedures. However, we shall not indulge in the solutions of the general unconstrained minimization problem. Rather, we shall present a procedure for finding the minimum of $S(\theta^1, \ldots, \theta^m)$. For simplicity, we shall only consider $F(x, \theta^1, \ldots, \theta^m)$ which does not contain complements of functions. Before we give the procedure, we need some preliminary definitions.

Definition 5.3.1. Let $F(x, \theta^1, \ldots, \theta^m)$ be an inexact function in terms of the linear function $h_1(x, \theta^1), \ldots, h_m(x, \theta^m)$. Then, for $1 \leq r \leq m$, the linear function $h_r(x, \theta^r)$ is said to be *active at a point p* if and only if

$$F(p, \theta^1, \ldots, \theta^m) = h_r(p, \theta^r).$$

For example, let

$$h_1(x, \theta^1) = x_1 + x_2$$

and

$$h_2(x, \theta^2) = -x_1 - x_2$$

where $+$ and $-$ are ordinary addition and subtraction, and

$$F(p, \theta^1, \theta^2) = h_1(x, \theta^1) + h_2(x, \theta^2).$$

Then h_1 is active at the point $p = (1,1)$ since

$$F(p, \theta^1, \theta^2) = \text{Max} \left[h_1(p, \theta^1), h_2(p, \theta^2) \right]$$

$$= \text{Max} (2, -2)$$

$$= 2$$

$$= h_1(p, \theta^1).$$

Definition 5.3.2. Given a finite set $X = \{x^1, \ldots, x^N\}$, an N-vector $a = (a_1, \ldots, a_N)$ is said to be an *active indicator* for X if a_i is an element in $\{0, 1, \ldots, m\}$ for $i = 1, \ldots, m$. An active indicator $a = (a_1, \ldots, a_N)$ is said to be defined by $h_1, \ldots, h_m$ if a_i is given by $a_i = r_i$, where r_i is the least integer such that h_{r_i} *is active at* x^i.

Definition 5.3.3. Given a finite set $X = \{x^1, \ldots, x^N\}$ and an active indicator $a = (a_1, \ldots, a_N)$ for X, the partition of X by a is a family of subsets of X, $X_1, \ldots, X_m$ such that X_i is defined by

$$X_i = \{x^j \in X \,|\, a_j = i\}, \quad i = 1, \ldots, m.$$

Definition 5.3.4. Given an active indicator a which is defined by a set of linear functions, the set defined by

$$R_a = \{(\theta^1, \ldots, \theta^m) \,|\, h_1, \ldots, h_m \text{ defines the active indicator } a, \text{ where } h_i = \theta^i \cdot x \text{ for } i = 1, \ldots, m\}$$

is called an *active region associated with a*. It should be noted that an active region is a convex polyhedral cone with its vertex at the origin since it can be also defined by a set of linear inequalities.

We now state an algorithm as follows.

Algorithm 5.3.1.
Step 1: Arbitrarily select an active indicator a^0.
Step 2: $k = 0$.
Step 3: Let $X_1^k, \ldots, X_m^k$ be the partition of X by a^k.
Step 4: Let Y_i^k be the set given by $Y_i^k = \{y \,|\, y = f(x) \text{ for } x \in X_i^k\}$, $i = 1, \ldots, m$.
Step 5: For $i = 1, \ldots, m$ find the linear regression for (Y_i^k, X_i^k), i.e. solve

$$(\tilde{X}_i^k)^T \tilde{X}_i^k (\theta^i)^k = (\tilde{X}_i^k)^T \tilde{Y}_i^k,$$

where $\tilde{X}_i^k$ and $\tilde{Y}_i^k$ are the matrices whose rows are elements of X_i^k and Y_i^k (T means transpose of a matrix).

Step 6: For $i = 1, \ldots, m$ if the equations defined in Step (5) have a solution, let $h_i^k = (\theta^i)^k \cdot x$; otherwise, let $h_i^k = h_i^{k-1}$.
Step 7: Find the active indicator b^k defined by $h_1^k, \ldots, h_m^k$.
Step 8: If $b_j^k = a_j^k$ for $j = 1, \ldots, N$, terminate the algorithm and output $(\theta^1)^k, \ldots, (\theta^m)^k$; otherwise, go to (9).
Step 9: Let $a^{k+1} = b^k$.
Step 10: Set $k = k + 1$ and go to (3).

The above algorithm is based upon the following theorem concerned with a property of the error sum of square $S(\theta^1, \ldots, \theta^m)$ as defined above.

Theorem 5.3.1. $S(\theta^1, \ldots, \theta^m)$ *is a piecewise quadratic function composed of a finite number of quadratic functions of* $\theta^1, \ldots, \theta^m$.

Proof. For each active indicator $a = (a_1, \ldots, a_N)$, let R_a be the active region associated with a. For each element in R_a,

$$S(\theta^1, \ldots, \theta^m) = \sum_{i=1}^{N} \left[y^i - F(x^i, \theta^1, \ldots, \theta^m) \right]^2$$
$$= \sum_{i=1}^{N} \left[y^i - h_{a_i}(x^i, \theta^{a_i}) \right]^2.$$

However, $\sum_{i=1}^{N} \left[y^i - h_{a_i}(x^i, \theta^{a_i}) \right]^2$ is a quadratic function; hence $S(\theta^1, \ldots, \theta^m)$ is a quadratic function in R_a. Since there is a finite number of active indicators, there is a finite number of active regions. Therefore, $S(\theta^1, \ldots, \theta^m)$ is a piecewise quadratic function composed of a finite number of quadratic functions of $(\theta^1, \ldots, \theta^m)$.   Q.E.D.

In Algorithm 5.3.1 the initial active indicator a^0 is used to calculate values $(\theta^i)^0$, $i = 1, \ldots, m$. For $k > 0$, a^k is defined by $h_1^k, \ldots, h_m^k$. For a^k, where $k > 0$,

$$S(\theta^1, \ldots, \theta^m) = \sum_{i=1}^{N} \left[y^i - h_{a_i^k}(x^i, \theta^{a_i^k}) \right]^2.$$

If a minimum solution of $S(\theta^1, \ldots, \theta^m)$ is in the active region R_{a^k}, then we may obtain it by setting the first derivatives of equation (5.3.1) to zero. This is equivalent to step 5 of Algorithm 5.3.1. This process is continued until Algorithm 5.3.1 terminates. It should be noted that for some initial active indicators, it is possible that Algorithm 5.3.1 may terminate at local minimum solutions. However, for all practical reasons, it is clear that the number of initial active indicators that can lead to the minimum solutions of $S(\theta^1, \ldots, \theta^m)$ is far greater than the number of initial active indicators that lead to non-minimum solutions.

5.4. DETECTION OF HAZARDS

5.4.1. Static Hazards

Any physical switching system has delays associated with gates and interconnecting lines. While these delays are not important in synchronous systems because of the clock pulse, their effects on the operation of asynchronous systems are significant. These delays are known as stray delays to distinguish them from delays that are inserted in the system to ensure proper operation. Stray delays in gates may result because the gates may operate only when their inputs have values above some threshold value, whereas stray delays in lines may represent the propagation time of signals along the lines. A system is said to contain a hazard if there exists some possible combination of values of stray delays which will produce a spurious pulse or cause the system to enter an incorrect static state for some input change. It is important to note that a hazard represents only a possibility of malfunction. A specific logic system may not malfunction even though a hazard exists, because the relative magnitude of actual stray delays may ensure proper operation of the system. The possibility of the occurrence of spurious pulses on the outputs of a purely combinational circuit for certain input changes is called a combinational hazard. Combinational hazards are usually divided into two classes—static and dynamic. A static hazard is said to be present when the output of a circuit is required to remain constant during a transition, but, for some distribution of stray delays, the output may contain one or more pulses (i.e. the output changes an even number of times). Static hazards are classified as 0- and 1-hazards, depending on whether the output is specified to be 0 or 1 during the transition. A dynamic hazard may produce a sequence of three (or a greater odd number) output changes when a single change is required.

The following presents necessary and sufficient conditions for the presence of static hazards in two-level combinational circuits for single input variable changes utilizing the concept of fuzzy logic.

Considering the hazardous system as a fuzzy structure enables us to treat the static hazard as an ill-defined structure which cannot admit of a precise pattern. The hazardous variable implies imprecision in the system which stems not from randomness but from a lack of sharp transition from membership in a class to non-membership in it. In what follows we shall present a technique to identify hazardous switching functions, focusing our attention on the fuzzy structure of the logic system during its transient behavior. Even though it may appear incongruous to mention logic design and fuzziness in the same breath, this incongruity becomes less paradoxical if we examine the nature of hazards in combinational systems. Indeed, it is quite possible that during the transient state the input signal has a fuzzy structure representation. Intuitively, fuzziness is a type of imprecision which stems from a grouping of elements into classes which do not have sharply defined boundaries, that is, in which there is no *sharp transition* from membership to non-membership. Thus the transition of a state has a fuzzy behavior during the transition time, since this is a member in an ordered set of operations, some of which are fuzzy in nature, e.g. "switch x is closed slightly at each unit of time until y is approximately open." Treating such operations in a precise mathematical way provides a means of designing combinational systems when the conventional non-fuzzy techniques become infeasible. The success of the fuzzy procedure to be described lies in its philosophical and mathematical description of the hazard in combinational systems.

Intuitively, a fuzzy n-variable function is a mapping on Z^n to Z, $Z = [0,1]$ and Z^n is the set of all fuzzy vectors $(x_1, x_2, \ldots, x_n)$, $x_i \in Z$, $i = 1, 2, \ldots, n$.

Definition 5.4.1. Let $x = (x_1, x_2, \ldots, x_n)$ and $y = (y_1, y_2, \ldots, y_n)$ be two binary n-dimensional vectors. Two binary n-dimensional vectors are adjacent iff they differ in exactly one component, i.e.

$$y = (x_1, x_2, \ldots, x_{j-1}, \bar{x}_j, x_{j+1}, \ldots, x_n).$$

Definition 5.4.2. Variable x_j is called a *perfect fuzzy variable* iff its grade membership $\mu_A(x_j)$ is a number in the semi-closed intervals $(0,1]$ or $[0,1)$.

Definition 5.4.3. The fuzzy transmission vector $T^y_{x_j} \in Z^n$ is defined as the transition vector from vector x to its adjacent vector y (i.e. $T^y_x = x$ such that x_j is a perfect fuzzy variable).

Definition 5.4.4. A B-fuzzy n-variable function $f(x)$, $x = (x_1, x_2, \ldots, x_n)$, is any fuzzy function over n variables which is either constantly 1 or 0, or obtained from its arguments $x_1, x_2, \ldots, x_n$ by successive application of Max, Min, and complement. The following is the key theorem to the algorithm, as will be shown later; its proof is a trivial extension of the proof in the ternary case given by Yoeli and Rinon.

Theorem 5.4.1. Let $f(x)$, $x = (x_1, x_2, \ldots, x_n)$ be a *B-fuzzy n-variable function and ξ, ρ any adjacent binary n-dimensional vectors. Then*

$$f(T^\rho_{\xi_j}) \neq f(\xi). \tag{5.4.1}$$

In a similar way we define now a V-fuzzy function as a fuzzy function $f(x)$, such that $f(\xi)$ is a binary function for every binary n-dimensional vector ξ. It is clear that a V-fuzzy function f induces a binary function F such that

$$F: \{0,1\}^n \rightarrow \{0,1\} \tag{5.4.2}$$

determined by $F(\xi) = f(\xi)$ for every binary n-dimensional vector ξ.

If the B-fuzzy function f describes the complete behavior of a binary combinational system, its steady-state behavior is represented by F, the binary function induced by f. Let $f(x)$ be an n-dimensional V-fuzzy function and ξ, ρ adjacent binary n-dimensional vectors. The vector $T_{\xi_j}^\rho$ is a *static hazard* of f iff $f(\xi) = f(\rho) \neq f(T_{\xi_j}^\rho)$.

If $f(\xi) = f(\rho) = 1$, $T_{\xi_j}^\rho$ is a 1-hazard.

If $f(\xi) = f(\rho) = 0$, $T_{\xi_j}^\rho$ is a 0-hazard.

If f is B-fuzzy and $T_{\xi_j}^\rho$ is a static hazard then $f(T_{\xi_j}^\rho)$ has a perfect fuzzy value, i.e. $f(T_{\xi_j}^\rho) \in [0,1)$ or $(0,1]$.

Definition 5.4.5. A combinational system is a static-hazard-free (SHF) system if and only if its B-fuzzy function f is SHF.

5.4.2. Hazard Detection By Fuzzy Presentation

Consider the static hazard as a malfunction represented by an actual or potential deviation from the intended behavior of the system. We can detect all static hazards of the V-fuzzy function $f(x)$ by considering the following extension of Shannon normal form.

Let $f(x)$, $x = (x_1, x_2, \ldots, x_n)$, be a fuzzy function and denote the vector $(x_1, x_2, \ldots, x_{j-1}, x_{j+1}, \ldots, x_n)$ by x^j. By successive applications of the rules of fuzzy algebra the function $f(x)$ may be expanded about, say, x_j as follows:

$$f(x) = x_j f_1(x^j) + \bar{x}_j f_2(x^j) + x_j \bar{x}_j f_3(x^j) + f_4(x^j) \tag{5.4.3}$$

where f_1, f_2, f_3 and f_4 are also fuzzy functions. It is clear that the same expansion holds when the fuzzy functions are replaced by B-fuzzy functions of the same dimension.

Let ξ and ρ be two adjacent n-dimensional binary vectors that differ only in their j^{th} component. Treating ξ_j as a perfect fuzzy variable during transition time implies that $T_{\xi_j}^\rho$ is a 1-hazard of f iff $f(\xi) = f(\rho) = 1$ and $f(T_{\xi_j}^\rho) \in [0,1)$.

We shall now show that the above conditions for the vector $T^\rho_{\xi_j}$ to be 1-hazard yield the following result.

Theorem 5.4.2. The vector $T^\rho_{\xi_j}$ is a 1-hazard of the B-fuzzy function $f(x)$ given by equation (5.4.3) iff the binary vector ξ^j is a solution of the following set of Boolean equations:

$$f_1(x^j) = 1, \quad f_2(x^j) = 1, \quad f_4(x^j) = 0.$$

Proof.
State 1: $\xi_j = 1$ and $\bar{\xi}_j = 0$ imply $f_1(\xi^j) + f_4(\xi^j) = 1$.
State 2: $\xi_j = 0$ and $\bar{\xi}_j = 1$ imply $f_2(\xi^j) + f_4(\xi^j) = 1$.
Transition state: $\xi_j \in (0,1)$ (which implies $\bar{\xi}_j \in (0,1)$), and thus

$$0 \leqslant \text{Max} \left\{ \text{Min} \left[\xi_j, f_1(\xi^j) \right], \text{Min} \left[\bar{\xi}_j, f_2(\xi^j) \right], \text{Min} \left[\xi_j, \bar{\xi}_j, f_3(\xi^j) \right], f_4(\xi^j) \right\}.$$

It is clear from the transition state that $f_4(\xi^j)$ cannot be equal to 1, and thus

$$f_4(\xi^j) = 0, \quad f_1(\xi^j) = f_2(\xi^j) = 1. \tag{5.4.4}$$

It should be noted that the Yoeli–Rinon procedure produces the same result by using a ternary evaluation technique, which *a priori* imposes the value ½ over the grade membership of ξ_j. All binary $(n-1)$-dimensional vectors satisfying these conditions will be obtained by the resolution principle discussed later.

Similarly, the vector $T^\rho_{\xi_j}$ is a 0-hazard iff $f(\xi) = f(\rho) = 0$ and $f(T^\rho_\xi) \in (0,1]$. These conditions imply the following.

State 1: $\xi_j = 1$ and $\bar{\xi}_j = 0$ imply $f_1(\xi^j) + f_4(\xi^j) = 0$.
State 2: $\xi_j = 0$ and $\bar{\xi}_j = 1$ imply $f_2(\xi^j) + f_4(\xi^j) = 0$.
Transition state: $\xi_j \in (0,1)$ implies

$$0 \leqslant \text{Max} \left\{ \text{Min} \left[\xi_j, f_1(\xi^j) \right], \text{Min} \left[\bar{\xi}_j, f_2(\xi^j) \right], \text{Min} \left[\xi_j, \bar{\xi}_j, f_3(\xi^j) \right], f_4(\xi^j) \right\} \leqslant 1.$$

These simultaneous conditions are equivalent to

$$f_1(\xi^j) = f_2(\xi^j) = f_4(\xi^j) = 0, \quad f_3(\xi^j) = 1.$$

Hence the following theorem can be stated.

Theorem 5.4.3. The vector $T^\rho_{\xi_j}$ is a 0-hazard of the B-fuzzy function *f(x) given by equation (5.4.3) iff the vector ξ^j is a solution of the following set of Boolean equations:*

$$f_1(x^j) = f_2(x^j) = f_4(x^j) = 0, \quad f_3(x^j) = 1.$$

Dual procedures can be obtained if one expands the B-fuzzy function $f(x)$ by the dual Shannon form about, say, x_j as

$$f(x) = \left[x_i + g_1(x^i) \right] \left[\bar{x}_i + g_2(x^i) \right] \left[x_i + \bar{x}_i + g_3(x^i) \right] \left[g_4(x^i) \right].$$

It should be noted that a logical formula (form) can be interpreted in several ways and many properties of logical forms hold regardless of whether variables in the forms are interpreted as binary or fuzzy variables. One such property is demonstrated by Theorems 5.4.2 and 5.4.3. Even though these theorems are not new, they have been derived using no *a priori* assumptions as to the behavior or value of the switch during transition. Since this behavior is an imprecise one, treating the binary variable as a perfect fuzzy variable while in transition is quite a novel approach that explains the dynamic structure of the switch from the philosophical point of view as well as its mathematical model by means of fuzzy algebra. Thus the main contributions of this section are not the results of Theorems 5.4.2 and 5.4.3, but the basic assumptions of imprecisions by which they have been proved and by which the philosophical behavior of the transition structure has been analyzed.

5.4.3. Using The Resolution Principle

In order to make practical use of Theorems 5.4.2 and 5.4.3 one has to use some algebraic means to solve a set of simultaneous Boolean equations. In this section we investigate the use of the resolution principle used in theorem proving.

It will be shown that the matrix notation for the set of input clauses presented by Yelowitz leads to algebraic treatment of the resolution principle and its application in hazard detection. In addition, the technique can be easily implemented on digital computers, thus making it simpler and more efficient for practical use. Resolution-based solution proceeds by the assuming that a static hazard exists and then trying to derive a contradiction.

If f is represented in disjunctive normal form (d.n.f.), it is sufficient to convert f_1, f_2, and f_4 to conjunctive normal form (c.n.f.) in order to detect a 1-hazard (f_1, f_2, f_3, and f_4 if detection of a 0-hazard is desired).

Each of these is a function of at most $n - 1$ variables and therefore we allocate a matrix $M(i,j)$ with $n - 1$ columns (one column for each variable) in which each row of the matrix corresponds to a clause in c.n.f. Within a row an entry of 1, -1, or 0 represents occurrence, negated occurrence, or non-occurrence of the corresponding variable. Two rows i_1 and i_2 clash in column j if

$$M(i_1,j) \times M(i_2,j) = -1.$$

Two rows i_1 and i_2 are resolvable if there is exactly one column j such that i_1 and i_2 clash in column j. The generation of the resolvent row i_3, given two resolvable rows i_1 and i_2 which clash in column j, can be done as follows:

$$M(i_3,k) = \begin{cases} 0 & \text{if } k = j \\ M(i_2,k) & \text{if } k \neq j \text{ and } M(i_2,k) = M(i_1,k) \\ M(i_2,k) + M(i_1,k) & \text{if } k \neq j \text{ and } M(i_2,k) \neq M(i_1,k). \end{cases}$$

With the generation of a row consisting entirely of zeros, the contradiction is established.

It should be noted that the problem of keeping down the number of irrelevant clauses generated is beyond the scope of this book. Methods to deal with this problem are discussed in the literature.

The functions $f_1, \ldots, f_4$ can be transformed under DeMorgan laws. Namely, if a function has a true (1) value it is sufficient to convert it to c.n.f.; if it has a false (0) value, represent its negation.

Example 5.4.1. Let

$$f(x_1,x_2,x_3,x_4) = x_1 x_2 \bar{x}_3 x_4 + \bar{x}_1 \bar{x}_2 \bar{x}_3 x_4 + x_2 x_3 \bar{x}_4.$$

Suppose we want to check for a 1-static-hazard generated by x_1:

$$f(x_1,x_2,x_3,x_4) = x_1 f_1(x_2,x_3,x_4) + \bar{x}_1 f_2(x_2,x_3,x_4)$$
$$+ x_1 \bar{x}_1 f_3(x_2,x_3,x_4) + f_4(x_2,x_3,x_4)$$

where

$$f_1(x_2,x_3,x_4) = x_2 \bar{x}_3 x_4, \quad f_2(x_2,x_3,x_4) = \bar{x}_2 \bar{x}_3 x_4$$
$$f_3(x_2,x_3,x_4) = 0, \quad f_4(x_2,x_3,x_4) = x_2 x_3 \bar{x}_4.$$

Assume a 1-hazard exists; namely $f_1 = f_2 = 1$, $f_4 = 0$.

		x_2	x_3	x_4
f_1	(1)	1	0	0
	(2)	0	-1	0
	(3)	0	0	1
f_2	(4)	-1	0	0
	(5)	0	-1	0
	(6)	0	0	1
f_4	(7)	-1	-1	1

Rows (1) and (4) are resolvable since they clash only in column x_2. The resolvent row consists entirely of zeros, and a contradiction is established. Therefore the given function is 1-static-hazard free as far as the variable x_1 is considered.

Example 5.4.2. Let

$$f(x_1,x_2,x_3,x_4) = x_1x_2x_3 + x_1\bar{x}_4 + \bar{x}_1\bar{x}_2\bar{x}_3\bar{x}_4 + \overline{\bar{x}_2\bar{x}_3x_4}.$$

Checking for 1-hazard generated by x_1, we obtain

$$f_1(x_2,x_3,x_4) = x_2x_3 + \bar{x}_4, \qquad f_2(x_2,x_3,x_4) = \bar{x}_2\bar{x}_3\bar{x}_4$$

$$f_4(x_2,x_3,x_4) = x_2 + x_3 + \bar{x}_4.$$

Under 1-hazard assumption, $f_1 = f_2 = 1, f_4 = 0$.

		x_2	x_3	x_4
f_1	(1)	1	0	-1
	(2)	0	1	-1
f_2	(3)	-1	0	0
	(4)	0	-1	0
	(5)	0	0	-1
f_3	(6)	-1	0	0
	(7)	0	-1	0
	(8)	0	0	1

Rows (5) and (8) are resolvable since they clash only in column x_4 and thus the resolvent row consists entirely of zeros. Therefore we have a contradiction to our assumption and the function has no 1-hazard generated by x_1.

Example 5.4.3. Let

$$f(x_1,x_2,x_3,x_4) = x_1x_2 + (x_1 + x_4)(\bar{x}_1 + \bar{x}_3)$$

and let x_1 be the variable in question. Assume a 1-static hazard exists. Namely,

$$f_1(x_2,x_3,x_4) = x_2 + \bar{x}_3 = 1, \qquad f_2(x_2,x_3,x_4) = x_4 = 1$$

$$f_4(x_2,x_3,x_4) = \bar{x}_3x_4 = 0.$$

	x_2	x_3	x_4
f_1 (1)	1	−1	0
f_2 (2)	0	0	1
f_4 (3)	0	1	−1

We may generate the following new rows of the matrix where the numbers in parentheses indicate the generators of the new row:

	x_2	x_3	x_4	
(4)	0	1	0	(2,3)
(5)	1	0	0	(1,4)

Clauses (1) and (3) are subsumed by clauses (5) and (4), respectively, and may therefore be deleted. Evidently, rows (2), (4), and (5) cannot yield the derivation of the zero clause, and thus a 1-static hazard caused by x_1 may exist.

The problems of determining whether or not digital systems operate correctly, and of ensuring their proper operation even when some of their elements are fuzzy, are of both practical concern and theoretical interest. In this book the fuzzy treatment of the transient behavior of switching systems is discussed and a diagnostic procedure for detecting static hazards is demonstrated. In the combinational system which demonstrates hazardous behavior it is possible for the output signals to behave in an unpredicted manner under certain input transitions, namely, the system belongs to a set of ill-defined systems, some of which cannot admit any precise analysis. Perhaps the major reason for the ineffectiveness of classical techniques in dealing with transient behavior in systems lies in their failure to come to grips with the issue of fuzziness. This is due to the fact that transient behavior implies imprecision in the system, which stems not from randomness but from a lack of sharp transition between members of the class of input states. It is the same type of imprecision which arises when one is dealing, for example, with the class of systems which are approximately linear and other systems and classes which admit the possibility of partial membership in them.

5.5. APPLICATIONS OF FUZZY MATRIX THEORY

5.5.1. Pattern Recognition

Almost all of the research on pattern recognition has examined the problem of recognizing an isolated pattern that is embedded in a homogeneous background. Typically, the pattern is further simplified in that it is resolved and projected onto a two-dimensional matrix which enormously reduces the amount of potential information in the input pattern. This is usually the case with cluster analysis which has the objective of classifying experimental data in a certain number of categories where the elements of each category should be as similar as possible and dissimilar from that of other categories. This implies the existence of a measure of distance or similarity between the elements to be classified. The number of such categories may be fixed beforehand or may be a consequence of some constraints imposed on them.

In what follows our attention will be focused primarily on exploring the mathematical properties of inexact matrices and using them in pattern recognition. In most classification problems subjective information plays an important role and thus our analysis is data free and analytic in nature; namely, no data set of particular patterns is assumed and we shall make no attempt in the present report to discuss its possible applications to related problem areas such as decision processes, system modeling, or approximation theory.

Let $x_1, x_2, \ldots, x_n$ be given elements of a fuzzy set A in a space $\Omega = \{x\}$. By *abstraction* on $x_1, x_2, \ldots, x_n$ is meant the identification of those properties of $x_1, x_2, \ldots, x_n$ which they have in common and which, in aggregate, define the inexact set A. Let $\mu_A(x_j)$ denote the value of the characteristic function of A of element x_j in Ω. The set $\{(x_j, \mu_A(x_j))\}_n$ will be called a collection of *samples* or observations from A. The notion of "similarity" is defined as a generalization of the notion of equivalence. More concretely, a similarity relation S is a fuzzy relation which is reflexive, symmetric, and transitive. Thus, let x_i and x_j be elements of Ω, and let $\mu_S(x_i, x_j)$ denote the grade of membership of the ordered pair (x_i, x_j) in S. Then S is a similarity relation in Ω if $\forall\, x, y, z \in \Omega$

$$\mu_S(x,x) = 1 \quad \text{(reflexive)} \tag{5.5.1}$$

$$\mu_S(x,y) = \mu_S(y,x) \quad \text{(symmetric)} \tag{5.5.2}$$

$$\mu_S(x,z) \geq \underset{y}{\text{Max}} \left[\text{Min} \left(\mu_S(x,y), \mu_S(y,z) \right) \right] \quad \text{(transitive)} \tag{5.5.3}$$

Definition 5.5.1. Let $\mu_S(x,y)$ be a proximity relation as defined above. Then $\mu_S^n(x,y)$ is called the *n-purlieus relation,* defined as

$$\mu_S^n(x,y) = \underset{x_1, x_2, \ldots, x_{n-1} \in \Omega}{\text{Sup}} \left\{ \text{Min} \left[\mu_S(x, x_1), \mu_S(x_1, x_2), \ldots, \mu_S(x_{n-1}, y) \right) \right\}$$

for $n = 2, 3, \ldots$, since clearly $\mu_S^1(x,y) = \mu_S(x,y)$. Clearly we have

$$0 \leq \mu_S(x,y) \leq \mu_S^2(x,y) \leq \ldots \leq \mu_S^n(x,y) \leq \mu_S^{n+1}(x,y) \leq \ldots \leq 1$$

since

$$\mu_S^{n+1}(x,y) = \underset{x_1, x_2, \ldots, x_n \in \Omega}{\text{Sup}} \left\{ \text{Min} \left[\mu_S(x, x_1), \ldots, \mu_S(x_{n-1}, x_n), \mu_S(x_n, y) \right] \right\}$$

$$\geq \underset{x_1, x_2, \ldots, x_{n-1} \in \Omega}{\text{Sup}} \left\{ \text{Min} \left[\mu_S(x, x_1), \ldots, \mu_S(x_{n-1}, y), \mu_S(y, y) \right] \right\}$$

$$= \underset{x_1, x_2, \ldots, x_{n-1} \in \Omega}{\text{Sup}} \left\{ \text{Min} \left[\mu_S(x, x_1), \ldots, \mu_S(x_{n-1}, y), 1 \right] \right\}$$

$$= \mu_S^n(x,y).$$

Clearly, the *n-purlieus relation* defined as

$$\mu_S^n(x,y) \triangleq \underset{x \in \Omega}{\text{Sup}} \left\{ \text{Min} \left[\mu_S(x, x_1), \ldots, \mu_S(x_{n-1}, y) \right] \right\}$$

where $\vec{x} = (x_1, x_2, \ldots, x_{n-1}) \in \Omega$ (Ω is the $(n-1)$-fold Cartesian product of X with itself where $x, y \in X$) implies that for all $x, y \in X$ and all $n \geq 1$

$$0 \leq \mu_S^n(x,y) \leq \mu_S^{n+1}(x,y) \leq 1.$$

Consequently $\lim\limits_{n \to \infty} \mu_S^n(x,y) \triangleq \bar{\mu}_S(x,y)$ exists by the monotone convergence principle; namely, for every $\epsilon > 0$, there is an integer N such that $\left| \mu_S^n(x,y) - \bar{\mu}_S(x,y) \right| < \epsilon$ for $n > N$. Since the sequence is non-decreasing and is bounded from above and below we can conclude that the limit exists.

Definition 5.5.2. Let x and y be two elements of Ω, and let $\mu_S^n(x,y)$ be the n-purlieus relation as defined above. Then we define the *propinquity* $\bar{\mu}(x,y)$ in $[0,1]$ such that

$$\bar{\mu}(x,y) = \lim_{n \to \infty} \mu_S^n(x,y).$$

Definition 5.5.3. Let $x,y \in \Omega$. Then x and y are said to have a *threshold* relation (xR_Ty) iff $\bar{\mu}_S(x,y) \geq T$. It is easy to show that $\forall\, x,y,z \in \Omega$ we have

$$\bar{\mu}_S(x,z) \geq \underset{y}{\text{Min}}\, \big[\bar{\mu}_S(x,y), \bar{\mu}_S(y,z)\big].$$

Theorem 5.5.1. The threshold relation is a similarity relation in Ω.

Proof.
(i) $xR_Tx \quad \forall T \in [0,1]$

since $1 = \mu_S(x,x) \leq \bar{\mu}_S(x,x) \leq 1 \quad \forall\, x \in \Omega$.

(ii) xR_Ty iff yR_Tx

since $\lim_{n \to \infty} \mu_S^n(x,y) = \bar{\mu}_S(x,y) = \bar{\mu}_S(y,x) = \lim_{n \to \infty} \mu_S^n(y,x).$

(iii) $xR_Ty \wedge yR_Tz \to xR_Tz$

since $\bar{\mu}_S(x,z) \geq \underset{y}{\text{Min}}\, \big[\bar{\mu}_S(x,y), \bar{\mu}_S(y,z)\big].$ Q.E.D.

Clearly we can associate with every relation an appropriate matrix to represent the relation, and hence we can classify the patterns using the partition induced by the threshold relation.

The fuzzy matrix approach is a pictorial representation of a graph-theoretic principle involving the selection of a threshold distance. Once the threshold distance d_0 is selected, two elements are said to be in the same cluster if the distance between them is less than d_0. This procedure can easily be generalized to apply to arbitrary similarity measures. Suppose that we pick a threshold value d_0 and say that α is similar to β if $s(\alpha,\beta) > d_0$.

$$S_{ij} = \begin{cases} 1 & \text{if } s(\alpha_i, \beta_j) > d_0 \\ 0 & \text{otherwise} \end{cases} \qquad i,j = 1, \ldots, n.$$

This matrix defines a *similarity graph* in which nodes correspond to points and an edge joins node i and node j if and only if $S_{ij} = 1$.

The clusterings produced by the single-linkage algorithm and by a modified version of the complete-linkage algorithm are readily described in terms of this graph. With the single-linkage algorithm two samples α and β are in the same cluster if and only if there exists a chain $\alpha_1, \alpha_2, \ldots, \alpha_k$ such that α is similar to α_1, α_1 is similar to α_2, and so on for the whole chain. Thus, this clustering corresponds to the connected components of the similarity graph. With the complete-linkage algorithm all samples in a given cluster must be similar to one another, and no sample can be in more than one cluster. If we drop this second requirement, then this clustering corresponds to the maximal complete subgraphs of the similarity graph, the "largest" subgraphs with edges joining all pairs of nodes. (In general, the clusters of the complete-linkage algorithm will be found among the maximal complete subgraphs, but they cannot be determined without knowing the unquantized similarity values.)

It is clear that if R_T is a threshold relation induced by $\mu_S(x,y)$ and R'_T is a threshold relation induced by $\mu'_S(x,y)$ and $\mu_S(x,y) \leq \mu'_S(x,y)$ for all $x, y \in \Omega$, then R_T refines R'_T.

It is our assumption that if $x \neq y$ then $\bar{\mu}_S(x,y) \in [0,1)$ and thus the function $\bar{\eta}_S(x,y) = 1 - \bar{\mu}_S(x,y)$ acts as a distance function. This is obvious, since

    (i) $\bar{\eta}_S(x,y) > 0$ for $x \neq y$ and $\bar{\eta}_S(x,x) = 0$.

    (ii) $\bar{\eta}_S(x,y) = \bar{\eta}_S(y,x)$

and

    (iii) $\bar{\eta}_S(x,z) \leq \bar{\eta}_S(x,y) + \bar{\eta}_S(y,z)$

as

$$\bar{\mu}_S(x,z) \geq \underset{y}{\text{Min}} \left[\bar{\mu}_S(x,y), \bar{\mu}_S(y,z) \right] \geq \bar{\mu}_S(x,y) + \bar{\mu}_S(y,z) - 1.$$

We shall assume for simplicity of the analysis that we deal with only a finite number of patterns, and hence we shall consider our threshold relation on finite sets only.

Theorem 5.5.2. Let $x_1, x_2, \ldots, x_n \in \Omega$ where n is a finite number. Then

$$R_T = \psi(R_T^*) = (R_T^*)^{n-1}$$

where

$$xR_T^*y \text{ iff } \mu_S(x,y) \geq T$$

and

$$\psi(R_T^*) = \underset{i}{\text{Sup}} \ (R_T^*)^i = \text{adj} \ (R_T^*).$$

Proof.

(i) Clearly $\psi(R_T^*)$ refines R_T since

$$\psi(R_T^*) = \underset{j}{\cup} \ (R_T^*)^j$$

and if $x\left[\psi(R_T^*)\right]y$ then $\exists \ x_1, x_2, \ldots, x_{n-1} \in \Omega$ such that

$$\mu_S(x, x_1) \geq T, \ \ldots, \ \mu_S(x_{n-1}, y) \geq T$$

and hence

$$\mu_S^n(x,y) \geq \text{Min} \ \left[\mu_S(x, x_1), \ldots, \mu_S(x_{n-1}, y)\right] \geq T$$

which implies that

$$\bar{\mu}_S(x,y) \geq \mu_S^n(x,y) \geq T$$

and thus xR_Ty.

(ii) Assume xR_Ty. then

$$\mu_S(x,y) = \mu_S^{n-1}(x,y)$$
$$= \underset{x_1, x_2, \ldots, x_{n-1} \in \Omega}{\text{Max}} \ \left\{\text{Min} \ \left[\mu_S(x, x_1), \ldots, \mu_S(x_{n-2}, y)\right]\right\} \geq T.$$

Therefore

$$\exists \ x_1, x_2, \ldots, x_{n-2} \in \Omega$$

such that

$$\mu_S(x, x_1) \geq T$$
$$\mu_S(x_1, x_2) \geq T$$
$$\vdots$$
$$\mu_S(x_{n-2}, y) \geq T$$

and thus we have

$$xR_T^*x_1, \ \ x_1R_T^*x_2, \ldots, \ \ x_{n-2}R_T^*y.$$
$$\rightarrow x(R_T^*)^{n-1}y \text{ exists}$$
$$\rightarrow x\left[\psi(R_T^*)\right]y \text{ exists.}$$

(iii) Let $R_T^* = [p_{ij}]$. The ij entry of $(R_T^*)^2$ is $\sum_{k=1}^n p_{ik} p_{kj}$, and this term has a grade membership of

$$\underset{k}{\text{Max}} \left[\text{Min} (p_{ik}, p_{kj}) \right]$$

iff there is a direct path between vertices i and j or there is a path from i to j through one intermediate vertex. Extending this argument to $(R_T^*)^i$ it is clear that no path requires more than $n - 2$ intermediate vertices, since there are only n vertices and internal loops are excluded. Hence, the ij entry of $(R_T^*)^{n-1}$ has a grade membership of

$$\underset{\text{subterms}}{\text{Max}} \left\{ ij \text{ terms of } (R_T^*)^{n-1} \right\}$$

iff i and j are connected, namely $(R_T^*)^{n-1} = \psi(R_T^*)$. Q.E.D.

Corollary 5.5.1.

$R_T^* = \psi(R_T^*) \leftrightarrow (R_T^*)^2 = R_T^*.$

Based on the above result, Algorithm 5.5.1 is presented to compute $\psi(R_T^*)$.

Algorithm 5.5.1. Given the matrix R_T^* constructed from the inexact patterns $x_1, x_2, \ldots, x_n$. Generate the matrix $(R_T^*)^l$, beginning with $l = 1$ and increasing l by 1, until $(R_T^*)^l = (R_T^*)^{l+1}$ for some l.

The repeated matrix multiplication makes Algorithm 5.5.1 unattractive from an efficiency viewpoint. Algorithm 5.5.2 achieves the same result and requires only a single scan over the matrix. In fact, Algorithm 5.5.2 works correctly on a wider range of input, since it is not required that the diagonal elements of the input matrix is 1, as with Algorithm 5.5.1.

Algorithm 5.5.2.

Step 1: Label vertices of R_T^* by the integers $1, \ldots, N$.

Step 2: Generate the matrix R_T^*.

Step 3: DO $K = 1$ TO N

Step 4: DO $I = 1$ TO N

Step 5: IF $R_T^*(I,K) \neq 0$ THEN

Step 6: DO $J = 1$ TO N

Step 7: $R_T^*(I,J) = \text{Max} \left[R_T^*(I,J), \text{Min} (R_T^*(I,K), R_T^*(K,J)) \right]$

Step 8: END

Step 9: END

Step 10: END

The basic idea is to scan down *column K,* and for each non-zero element encountered (say in row *I*), each element in *row I* (say element $R_T^*(I,J)$) is possibly improved by comparing $R_T^*(I,J)$ with Min $(R_T^*(I,K), R_T^*(K,J))$. A rigorous proof of correctness is achieved by attaching the following inductive assertion *W* between steps (7) and (8):

(W) $R_T^*(I,J) = G(I,J,K)$

where

$G(I,J,K)$ = Max {Min (all chains from *I* to *J* such that each intermediate element has a label $\leqslant K$)}.

Before proving that assertion *W* is true whenever control leaves step (7), it is noted that the relation

$R_T^*(I,J) = G(I,J,N)$

is the desired relation at the termination of the algorithm, since $G(I,J,N)$ = Max {Min (all chains from *I* to *J*)}. Assertion *W* is proved by induction on *K*.

(i) $K = 1$. The first time *W* is reached *K* has the value 1, and analysis shows that

$$R_T^*(I,J) = \text{Max}\left[R_{T_0}^*(I,J), \text{Min}\left(R_{T_0}^*(I,1), R_{T_0}(1,K)\right)\right]$$

where $R_{T_0}^*$ represents the original matrix, and the right-hand side of the equation is $G(I,J,1)$.

(ii) Assume $R_T^*(I,J) = G(I,J,K)$, $1 \leqslant K < N$. Then we have to show $R_T^*(I,J) = G(I,J,K + 1)$.

There are two subcases to consider. If $G(I,J,K + 1)$ does not involve element $K + 1$, then no change is made to the matrix and the desired result is true. If $G(I,J,K + 1)$ does involve element $K + 1$, then we can guarantee that element $K + 1$ appears only once, since loops do not increase the Max of any chain. Thus we can break the optimal chain into two subchains, $R_T^*(I,K + 1)$ and $R_T^*(K + 1,J)$. Since both subchains involve intermediate elements numbered $\leqslant K$, the inductive hypothesis applies to each subchain and the desired result follows.

It is interesting to note that during the process of computing the characteristic fuzzy matrix, minimization of the fuzzy structures are possible. In general, one cannot apply binary minimization techniques and thus more specific methods directed toward the minimization of fuzzy functions should be developed. Generally, an indexed collection

$$\lambda = \{D_\alpha; \alpha \in J\}$$

of subsets of a set S satisfying

$$S = \bigcup_{\alpha \in J} D_\alpha$$

$$D_\beta \cap D_\gamma = \varnothing \quad \text{for all } \beta \neq \gamma \in J$$

is said to be a *partition* of S.

Clearly we can apply the notions discussed above for the partition of our pattern set into disjoint classes which depends on the *a priori* assigned threshold.

Example 5.5.1. Let $X = \{x_1, x_2, x_3, x_4\}$ and $\mu_S(x_i, x_j)$ $i,j = 1, 2, 3, 4$ be as follows.

$$R_T^* = \begin{array}{c} \\ x_1 \\ x_2 \\ x_3 \\ x_4 \end{array} \begin{array}{cccc} x_1 & x_2 & x_3 & x_4 \\ \begin{bmatrix} 1 & 0 & T_2 & T_1 \\ 0 & 1 & T_4 & T_3 \\ T_2 & T_4 & 1 & \overline{T_1} \\ \overline{T_1} & T_3 & \overline{T_1} & 1 \end{bmatrix} \end{array}$$

Let

$$T_1 = 0.3$$

$$T_2 = 0.5$$

$$T_3 = 0.6$$

$$T_4 = 0.9$$

Then

$$\psi(R_T^*) = \begin{array}{c} \\ x_1 \\ x_2 \\ x_3 \\ x_4 \end{array} \begin{array}{cccc} x_1 & x_2 & x_3 & x_4 \\ \begin{bmatrix} 1 & 0.5 & 0.5 & \overline{0.5} \\ 0.5 & 1 & 0.9 & 0.3 \\ 0.5 & 0.9 & 1 & 0.3 \\ 0.5 & 0.3 & 0.3 & 1 \end{bmatrix} \end{array}$$

and we have the partitions

$$R_{T=1} = \{[x_1], [x_2], [x_3], [x_4]\}$$
$$R_{1>T>0.5} = \{[x_1], [x_2,x_3], [x_4]\}$$
$$R_{0.5 \geqslant T>0.3} = \{[x_1,x_2,x_3], [x_4]\}$$
$$R_{0.3 \geqslant T \geqslant 0} = \{[x_1,x_2,x_3,x_4]\}.$$

Example 5.5.2.

$$R_I^* = \begin{array}{c} \\ x_1 \\ x_2 \\ x_3 \\ x_4 \\ x_5 \end{array} \begin{array}{ccccc} x_1 & x_2 & x_3 & x_4 & x_5 \\ \hline 1 & 0.8 & 0 & 0.1 & 0.2 \\ 0.8 & 1 & 0.4 & 0 & 0.9 \\ 0 & 0.4 & 1 & 0 & 0 \\ 0.1 & 0 & 0 & 1 & 0.5 \\ 0.2 & 0.9 & 0 & 0.5 & 1 \end{array}$$

$$\psi(R_I^*) = \begin{array}{c} \\ x_1 \\ x_2 \\ x_3 \\ x_4 \\ x_5 \end{array} \begin{array}{ccccc} x_1 & x_2 & x_3 & x_4 & x_5 \\ \hline 1 & 0.8 & 0.4 & 0.5 & 0.8 \\ 0.8 & 1 & 0.4 & 0.5 & 0.9 \\ 0.4 & 0.4 & 1 & 0.4 & 0.4 \\ 0.5 & 0.5 & 0.4 & 1 & 0.5 \\ 0.8 & 0.9 & 0.4 & 0.5 & 1 \end{array}$$

Thus we have the partitions

$$R_{T=1} = \{[x_1], [x_2], [x_3], [x_4], [x_5]\}$$
$$R_{1>T>0.8} = \{[x_1], [x_2,x_5], [x_3], [x_4]\}$$
$$R_{0.8 \geqslant T>0.5} = \{[x_1,x_2,x_5], [x_3], [x_4]\}$$
$$R_{0.5 \geqslant T>0.4} = \{[x_1,x_2,x_4,x_5], [x_3]\}$$
$$R_{0.4 \geqslant T \geqslant 0} = \{[x_1,x_2,x_3,x_4,x_5]\} .$$

Example 5.5.3.

Let $\mu_S(x,y) = \dfrac{1}{1 + |x - y|}$, $x,y \in N$

$$R_I^* = $$

	0	1	2	3	4	5	6	7	8	...
0	1	$\frac{1}{2}$	$\frac{1}{3}$	$\frac{1}{4}$	$\frac{1}{5}$	$\frac{1}{6}$	$\frac{1}{7}$	$\frac{1}{8}$	$\frac{1}{9}$	...
1	$\frac{1}{2}$	1	$\frac{1}{2}$	$\frac{1}{3}$	$\frac{1}{4}$	$\frac{1}{5}$	$\frac{1}{6}$	$\frac{1}{7}$	$\frac{1}{8}$	...
2	$\frac{1}{3}$	$\frac{1}{2}$	1	$\frac{1}{2}$	$\frac{1}{3}$	$\frac{1}{4}$	$\frac{1}{5}$	$\frac{1}{6}$	$\frac{1}{7}$	...
3	$\frac{1}{4}$	$\frac{1}{3}$	$\frac{1}{2}$	1	$\frac{1}{2}$	$\frac{1}{3}$	$\frac{1}{4}$	$\frac{1}{5}$	$\frac{1}{6}$	...
4	$\frac{1}{5}$	$\frac{1}{4}$	$\frac{1}{3}$	$\frac{1}{2}$	1	$\frac{1}{2}$	$\frac{1}{3}$	$\frac{1}{4}$	$\frac{1}{5}$	...
5	$\frac{1}{6}$	$\frac{1}{5}$	$\frac{1}{4}$	$\frac{1}{3}$	$\frac{1}{2}$	1	$\frac{1}{2}$	$\frac{1}{3}$	$\frac{1}{4}$	...
6	$\frac{1}{7}$	$\frac{1}{6}$	$\frac{1}{5}$	$\frac{1}{4}$	$\frac{1}{3}$	$\frac{1}{2}$	1	$\frac{1}{2}$	$\frac{1}{3}$	...
7	$\frac{1}{8}$	$\frac{1}{7}$	$\frac{1}{6}$	$\frac{1}{5}$	$\frac{1}{4}$	$\frac{1}{3}$	$\frac{1}{2}$	1	$\frac{1}{2}$	...
8	$\frac{1}{9}$	$\frac{1}{8}$	$\frac{1}{7}$	$\frac{1}{6}$	$\frac{1}{5}$	$\frac{1}{4}$	$\frac{1}{3}$	$\frac{1}{2}$	1	...
⋮										⋱

$$\psi(R_I^*) = $$

	0	1	2	3	4	5	6	7	8	...
0	1	$\frac{1}{2}$	$\frac{1}{2}$	$\frac{1}{2}$	$\frac{1}{2}$	$\frac{1}{2}$	$\frac{1}{2}$	$\frac{1}{2}$	$\frac{1}{2}$	...
1	$\frac{1}{2}$	1	$\frac{1}{2}$	$\frac{1}{2}$	$\frac{1}{2}$	$\frac{1}{2}$	$\frac{1}{2}$	$\frac{1}{2}$	$\frac{1}{2}$	...
2	$\frac{1}{2}$	$\frac{1}{2}$	1	$\frac{1}{2}$	$\frac{1}{2}$	$\frac{1}{2}$	$\frac{1}{2}$	$\frac{1}{2}$	$\frac{1}{2}$	...
3	$\frac{1}{2}$	$\frac{1}{2}$	$\frac{1}{2}$	1	$\frac{1}{2}$	$\frac{1}{2}$	$\frac{1}{2}$	$\frac{1}{2}$	$\frac{1}{2}$	...
4	$\frac{1}{2}$	$\frac{1}{2}$	$\frac{1}{2}$	$\frac{1}{2}$	1	$\frac{1}{2}$	$\frac{1}{2}$	$\frac{1}{2}$	$\frac{1}{2}$	...
5	$\frac{1}{2}$	$\frac{1}{2}$	$\frac{1}{2}$	$\frac{1}{2}$	$\frac{1}{2}$	1	$\frac{1}{2}$	$\frac{1}{2}$	$\frac{1}{2}$	...
6	$\frac{1}{2}$	$\frac{1}{2}$	$\frac{1}{2}$	$\frac{1}{2}$	$\frac{1}{2}$	$\frac{1}{2}$	1	$\frac{1}{2}$	$\frac{1}{2}$	...
7	$\frac{1}{2}$	$\frac{1}{2}$	$\frac{1}{2}$	$\frac{1}{2}$	$\frac{1}{2}$	$\frac{1}{2}$	$\frac{1}{2}$	1	$\frac{1}{2}$	...
8	$\frac{1}{2}$	$\frac{1}{2}$	$\frac{1}{2}$	$\frac{1}{2}$	$\frac{1}{2}$	$\frac{1}{2}$	$\frac{1}{2}$	$\frac{1}{2}$	1	...
⋮										⋱

Clearly for $1 \geqslant T_1 \geqslant T_2 \geqslant 0$, R_{T_1} refines R_{T_2}; therefore, for every monotone non-increasing finite sequence of thresholds

$$0 \leqslant T_j \leqslant T_{j-1} \leqslant \ldots \leqslant T_2 \leqslant T_1 \leqslant 1$$

we can obtain a corresponding j-level hierarchy of clusters

$$J_i = \{\text{equivalence classes of } R_{T_i} \text{ in } X \,|\, 1 \leqslant i \leqslant j\}.$$

For each i, J_i is a partition of X and every class in J_{i+1} is the union of some non-empty class of subsets in J_i. It is interesting to note that if we define recursively

$$\Gamma_1 = R_{\hat{I}}^*$$

$$\Gamma_n = \Gamma_{n-1}^2$$

then

$$\Gamma_j = (R_{\hat{I}}^*)^{2^j}$$

and since by Theorem 5.5.2

$$\psi(R_{\hat{I}}^*) = (R_{\hat{I}}^*)^{n-1}$$

it is clear that

$$2^j \geqslant n - 1 \text{ will determine } \Gamma_j\big[\bar{\mu}_S(x,y)\big]$$

or

$$j \geqslant \log_2(n-1) \quad \text{will determine } \Gamma_j\big[\bar{\mu}_S(x,y)\big].$$

It should be noted that this is not a necessary condition since we can find certain cases for smaller J's to imply

$$\Gamma_j\big[\bar{\mu}_S(x,y)\big].$$

This hierarchical approach has been used by researchers using subjective information who performed experiments involving classification of static patterns such as portraits or dynamic patterns such as tropical cyclones. We shall not discuss these experiments; the interested reader, however, can refer to the papers reporting the result of these experiments given in the bibliography.

The characteristic fuzzy matrix represents a means by which the analysis of any finite fuzzy system can be obtained. The analysis technique that has been given is quite general and the use of matrix techniques leads to efficient computations, particularly in the description of fuzzy sequential procedures, such as decision making, and procedures involving sequences of imprecise operations which can be best represented by graphs and fuzzy chains.

Since systems which are either ill-defined or describe transitional behavior do not have a precise quantitative analysis, some graphical approach to represent these sytems is needed. It is in this sense that fuzzy logic analysis, through the use of fuzzy chains, might enable us to process decision-relevant information by using approximate relations to a primary set of precise data. This approach might be of use in areas such as decision processes, linguistics, sequential systems analysis, system-modeling approximation, and many more.

5.5.2. Role Theory

Role theory is an area of the social sciences which deals with the modeling of the interrelationships among individuals, positions, and tasks in an organization. In addition to the *formal* relations of the organizational structure, it is known that *informal* relations among individuals are very influential and often the dominant factor in performance. Thomason and Marinos propose the use of fuzzy sets as a natural analytical tool for describing these informal organizational relations. In what follows an illustration demonstrating the efficiency of fuzzy matrices in modeling organizational structures is presented.

To a first approximation an organization may be defined as a four-tuple of finite sets (S,P,T,R) where

$S = \{s_1,\ldots,s_m\}$ is the set of all individuals in the organization

$P = \{p_1,\ldots,p_n\}$ is the set of all positions in the organization

$T = \{t_1,\ldots,t_k\}$ is the set of all tasks to be accomplished by the organization within a specified time

$R = \{R_0,\ldots,R_j\}$ is the set of relationships between elements of S, P, and T.

It is interesting to note that this discrete-parameter model of an organization is time invariant, thus rendering, for example, the relationships in set R independent of time.

Among formal relations in R which affect decision making and performance of the organization are the following.

$R_j: P \times P \to \{0,1\}$ is the chain of command in the organization chart.

$R_1(p_i,p_j) = 1$ means p_i is an immediate superior of p_j.

$R_1(p_i,p_j) = 0$ means p_i is not an immediate superior of p_j, although p_i may control p_j via intermediate positions.

$R_2: T \times T \to \{0,1\}$ is the task-precedence relation which defines a relative ordering of tasks in time.

$R_3: S \times P \to \{0,1\}$ is the assignment of individuals to positions. Normally, R_3 defines a one-to-one correspondence of individuals to positions; however, if the organization is undermanned, an individual may occupy more than one position.

$R_4: P \times T \to \{0,1\}$ is the task-allocation relation which associates positions with the tasks for which they are directly responsible.

These formal relationships are conventional binary relations which define arcs with value 0 or 1 in a directed graph whose nodes are elements of the finite sets S, P, and T. A subsection of such a graph is shown in Fig. 5.7, which indicates that "individual h_i occupies position p_k (i.e. $R_3(s_i,p_k) = 1$) which is subordinate to p_j (i.e. $R_1(p_j,p_k) = 1$) but superior to p_l and p_m (i.e. $R_1(p_k,p_l) = R_1(p_k,p_m) = 1$) and directly responsible for tasks t_n and t_g (i.e. $R_4(p_k,t_n) = R_4(p_k,t_g) = 1$)".

It is well known that an informal social structure which exists outside these formal relationships exerts a strong influence on the functioning of an organization. Such loosely specified relations as "likes", "advises", and "communicates with", result in lines of influence and contact which the formal relations do not show. Unlike the clearly defined formal organizational structure given by R_1, R_2, R_3, and R_4, these informal relationships are ill defined and therefore best described as fuzzy relations.

We derive a fuzzy relation R_0 as a mapping from $S \times S$ into the closed interval $[0,1]$ such that $R_0(s_i,s_j)$ represents the degree to which s_i is R_0-related to s_j. For example, $R_0(s_i, s_j)$ may represent the degree to which s_i exerts informal but direct influence in s_j; thus, $R_0(s_i,s_j) = 0.9$ would mean s_i strongly influences s_j while $R_0(s_i,s_j) = 0.2$ would mean s_i has little direct influence on s_j.

In such relations there is indeterminacy due to vagueness rather than randomness; two individuals asked to evaluate the same relation

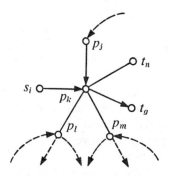

FIG. 5.7. Directed graph of formal relations.

would usually make different numerical assignments of connection values owing to the ambiguity in a fuzzy relationship. Only in the formal organizational relations are all the connections specified as present ($= 1$) or totally absent ($= 0$). However, in fuzzy relations the actual connection values generally are less critical than the differences between these values. Thus, it is more important to observe that

$$[R_0(s_i, s_j) = 0.91, R_0(s_j, s_i) = 0.23]$$

denotes the same trend as

$$[R_0(s_i, s_j) = 0.93, R_0(s_j, s_i) = 0.17]$$

than it is to emphasize that the values are not identical.

A subsection of a directed graph incorporating fuzzy relations is shown in Fig. 5.8. This graph illustrates that individual s_a influences position p_k (and hence influences positions and tasks placed under p_k by R_1 and R_4) even though position p_b held by s_a has no formal connection to p_k.

The strength of the connection of an element of S to an element of S, P, or T (i.e. the strength of a directed path, possibly through intermediate nodes) or a graph as in Fig. 5.7 is of paramount importance. The intrusion of informal relationships into the formal structure results in connections which are not binary valued $\{0,1\}$ but rather lie anywhere in the closed interval $[0,1]$. In determining the strength of a connection one must use the Max (Min) operation which asserts that the strength of the connection from one node to a second node on the graph is the maximum (Max) value of any directed path from the first node to the second, where each path takes the value of its weakest (Min) arc. Thus in Fig. 5.8 the strength of the connection of s_a to p_c is 1 since

$$\text{Min } [R_3(s_a, p_b), R_1(p_b, p_c)] = \text{Min } [1,1] = 1$$

while the strength of the connection of s_a to p_k is $R_0(s_a, s_i)$ since

$$\text{Min } [R_0(s_a, s_i), R_3(s_i, p_k)] = \text{Min } [R_0(s_a, s_i), 1]$$
$$= R_0(s_a, s_i).$$

The use of fuzzy matrices to represent the formal and informal relations defined previously facilitates the evaluation of nodal dependencies when the Max (Min) operation is to be used. As a quantitative example, consider the following relations for an organization in which $S = \{s_1, \ldots, s_4\}$ and $P = \{p_1, \ldots, p_5\}$.

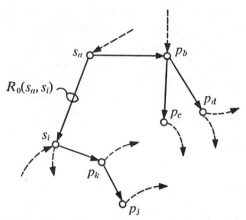

FIG. 5.8. Directed graph of formal and informal relations.

Influence Relation. $R_0: S \times S \to [0,1]$.

$$R_0 = \begin{bmatrix} 1 & 0.2 & 0.3 & 0.8 \\ 0.1 & 1 & 0.6 & 0.5 \\ 0.2 & 0.9 & 1 & 0.4 \\ 0.3 & 0.5 & 0.7 & 1 \end{bmatrix}$$

where, for instance, $R_0(s_2,s_3) = 0.6$ measures the informal but direct influence of s_2 on s_3. Note that $R_0(s_3,s_2) \neq 0.6$.

Superior–Subordinate Relation. $R_j: P \times P \to \{0,1\}$.

$$R_1 = \begin{bmatrix} 1 & 1 & 1 & 0 & 0 \\ 0 & 1 & 0 & 0 & 0 \\ 0 & 0 & 1 & 1 & 1 \\ 0 & 0 & 0 & 1 & 0 \\ 0 & 0 & 0 & 0 & 1 \end{bmatrix}$$

where, for instance, $R_1(p_1,p_2) = R_1(p_1,p_3) = 1$ means p_2 and p_3 are directly subordinate to p_1. Note $R_1(p_i,p_j) = 1$ which is in variance with other conventions. R_1 is a quantitative representation of the organization chart.

Personnel Assignment Relation. $R_3: S \times P \to \{0,1\}$.

$$R_3 = \begin{bmatrix} 0 & 1 & 0 & 0 & 0 \\ 0 & 0 & 1 & 1 & 0 \\ 1 & 0 & 0 & 0 & 0 \\ 0 & 0 & 0 & 0 & 1 \end{bmatrix}$$

where, for instance, $R_3(s_2,p_3) = R_3(s_2,p_4) = 1$ indicates that individual s_2 is occupying two positions, p_3 and p_4.

We wish to examine the influence of individuals on positions via formal *and* informal connections. Define a matrix M^k recursively as follows:

$$M_1 = R_3 \tag{5.5.4}$$

$$M^k = (R_0 \cdot M^{k-1}) \cup (M^{k-1} \cdot R_1) \tag{5.5.5}$$

where $\cdot$ is Max (Min) matrix multiplication in which sum and product become, respectively, Max and Min operations and $\cup$ is the term-wise union (Max) of matrices. The initial association of individuals with positions M^1 is given by the formal relation R_3; recursively, the informal influences defined by R_0 propagate via the formal chain of command R_1 so that individuals begin to affect positions other than their own. Using equations (5.5.4) and (5.5.5) one obtains

$$M^2 = \begin{bmatrix} 1 & 0.2 & 0.3 & 0.8 \\ 0.1 & 1 & 0.6 & 0.5 \\ 0.2 & 0.9 & 1 & 0.4 \\ 0.3 & 0.5 & 0.7 & 1 \end{bmatrix} \cdot \begin{bmatrix} 0 & 1 & 0 & 0 & 0 \\ 0 & 0 & 1 & 1 & 0 \\ 1 & 0 & 0 & 0 & 0 \\ 0 & 0 & 0 & 0 & 0 \end{bmatrix}$$

$$\cup \begin{bmatrix} 0 & 1 & 0 & 0 & 0 \\ 0 & 0 & 1 & 1 & 0 \\ 1 & 0 & 0 & 0 & 0 \\ 0 & 0 & 0 & 0 & 0 \end{bmatrix} \cdot \begin{bmatrix} 1 & 1 & 1 & 0 & 0 \\ 0 & 1 & 0 & 0 & 0 \\ 0 & 0 & 1 & 1 & 1 \\ 0 & 0 & 0 & 1 & 0 \\ 0 & 0 & 0 & 0 & 1 \end{bmatrix}$$

$$= \begin{bmatrix} 0.3 & 1 & 0.2 & 0.2 & 0.8 \\ 0.6 & 0.1 & 1 & 1 & 0.5 \\ 1 & 0.2 & 0.9 & 0.9 & 0.4 \\ 0.7 & 0.3 & 0.5 & 0.5 & 1 \end{bmatrix} \cup \begin{bmatrix} 0 & 1 & 0 & 0 & 0 \\ 0 & 0 & 1 & 1 & 1 \\ 1 & 1 & 1 & 0 & 0 \\ 0 & 0 & 0 & 0 & 1 \end{bmatrix}$$

$$= \begin{bmatrix} 0.3 & 1 & 0.2 & 0.2 & 0.8 \\ 0.6 & 0.1 & 1 & 1 & 1 \\ 1 & 1 & 1 & 0.9 & 0.4 \\ 0.7 & 0.3 & 0.5 & 0.5 & 1 \end{bmatrix}$$

$$M^3 = \begin{bmatrix} 0.7 & 1 & 0.5 & 0.5 & 0.8 \\ 0.6 & 0.6 & 1 & 1 & 1 \\ 1 & 1 & 1 & 0.9 & 0.9 \\ 0.7 & 0.7 & 0.7 & 0.7 & 1 \end{bmatrix} \cup \begin{bmatrix} 0.3 & 1 & 0.3 & 0.2 & 0.8 \\ 0.6 & 0.6 & 1 & 1 & 1 \\ 1 & 1 & 1 & 1 & 1 \\ 0.7 & 0.7 & 0.7 & 0.5 & 1 \end{bmatrix}$$

$$= \begin{bmatrix} 0.7 & 1 & 0.5 & 0.5 & 0.8 \\ 0.6 & 0.6 & 1 & 1 & 1 \\ 1 & 1 & 1 & 1 & 1 \\ 0.7 & 0.7 & 0.7 & 0.7 & 1 \end{bmatrix}$$

$$M^4 = \begin{bmatrix} 0.7 & 1 & 0.7 & 0.7 & 0.8 \\ 0.6 & 0.6 & 1 & 1 & 1 \\ 1 & 1 & 1 & 1 & 1 \\ 0.7 & 0.7 & 0.7 & 0.7 & 1 \end{bmatrix} \cup \begin{bmatrix} 0.7 & 1 & 0.7 & 0.5 & 0.8 \\ 0.6 & 0.6 & 1 & 1 & 1 \\ 1 & 1 & 1 & 1 & 1 \\ 0.7 & 0.7 & 0.7 & 0.7 & 1 \end{bmatrix}$$

$$= \begin{bmatrix} 0.7 & 1 & 0.7 & 0.7 & 0.8 \\ 0.6 & 0.6 & 1 & 1 & 1 \\ 1 & 1 & 1 & 1 & 1 \\ 0.7 & 0.7 & 0.7 & 0.7 & 1 \end{bmatrix} .$$

Clearly $M^4 = M^{4+k}$ for $k = 0,1,2,\ldots$, yielding the convergent limit matrix.

As shown in the previous discussion of fuzzy matrices, the powers of a fuzzy matrix will either converge to an idempotent limit matrix or oscillate with finite period. In the case of an $m \times n$ matrix M^k defined by equations 5.5.4 and 5.5.5 it can be shown that convergence will occur within $m + n - 1$ steps if R_0 and R_1 have 1's on the diagonal (see Chapter 4).

Fuzzy relations and their corresponding fuzzy matrices appear to be effective analytical tools for more detailed models of organizational structures. There are presently certain empirical axioms concerning those informal relations which must be defined and quantified in order to yield reasonable models; one such axiom is the notion that "informal relations will develop between individuals occupying positions responsible for the same task(s) due to the collaboration required." Informal relations do not necessarily have a positive effect on the functioning of an organization; for example, the relationship "dislikes" can be very detrimental.

The formulation of recursive schemes, similar to the above, which result in more complicated interaction of powers of fuzzy matrices describing organizational structure are entirely feasible.

REFERENCES

Fuzzy Neural Networks

1. McCulloch, W. S., and Pitts, W. A logical calculus of the ideas imminent in nervous activity", *Bull. Math. Biophys.* **5**, 1943.

2. von Neumann, J. Probabilistic logics and the synthesis of reliable organisms from unreliable components. *Automata Studies (Annals of Mathematics Studies,* No. 34), pp. 43–98, Princeton University Press, Princeton, N.J., 1956.

3. Moore, E. F. Gedanken-experiments on sequential machines, *Automata Studies (Annals of Mathematics Studies,* No. 34), pp. 129–153, Princeton University Press, Princeton, N.J., 1956.

4. Kleene, S. C., Representation of events in nerve nets and finite automata, *Automata Studies (Annals of Mathematics Studies,* No. 34), Princeton University Press, Princeton, N.J., 1956.

5. Rabin, M. O., and Scott, D. Finite automata and their decision problems. *IBM J. Res. Dev.,* **3**, 114–25, 1959.

6. McNaughton, R., and Yamada, H. Regular expressions and state graphs for automata. *Trans. IRE Electron. Computers* **EC-9** (1), 39–47, March 1960.

7. McNaughton, R. The theory of automata, a survey *Adv. Computers* **2**, 379–421, Academic Press, New York, 1961.

8. Gill, A. *Introduction to the Theory of Finite-State Machines,* McGraw-Hill, New York, 1962.

9. Harrison, M. A. *Introduction to Switching and Automata Theory.* McGraw-Hill, New York, 1965.

10. Harmanis, J., and Stearns, R. E. *Algebraic Structure Theory of Sequential Machines.* Prentice-Hall, Englewood Cliffs, N.J., 1966.

11. Ginsburg, S. *An Introduction to Mathematical Machine Theory.* Addison-Wesley, Reading, Mass., 1962.

12. Ginzburg, A. *Algebraic Theory of Automata.* Academic Press, New York, 1968.

13. Zadeh, L. A. Fuzzy Sets. *Inform. Control* **8,** 338–353, June 1965.

14. Zadeh, L. A. Biological application of the theory of fuzzy sets and systems. *Proc. Symp. on the Biocybernetics of the Central Nervous System,* 1968.

15. Wee, W. G., and Fu, K. S. A Formulation of Fuzzy Automata and Its Application as a Model of Learning Systems. *IEEE Trans. Syst. Sci. Cybernetics,* **SSC-5** (3), July 1969.

16. Mizumoto, M., Toyoda, J., and Tanaka, K. Some Considerations on Fuzzy Automata. *J. Computer Sys. Sci.* **3,** 409–22, 1969.

17. Santos, E. S. Maximin Automata. *Infor. Control,* **13,** 363–77, 1968.

18. Minsky, M. L. *Computation—Finite and Infinite Machines,* Prentice Hall, Englewood Cliffs, N.J., 1967.

19. Lee, E. T., and Zadeh, L. A., Note on fuzzy languages. *Inform. Sci.* **1,** 421–34, 1969.

20. Bellman, R. E., Kalaba, R., and Zadeh, L. A. Abstraction and pattern classification, *J. Math. Analysis Applic.* **13,** 1–7, 1966.

21. Zadeh, L. A. Fuzzy algorithms. *Inform. Control* **12,** 99–102, Feb. 1968.

22. Santos, E. S. Fuzzy algorithms. *Inform. Control* **17,** 326–39, Nov. 1970.

23. Bellman, R. E., and Zadeh, L. A. Decision-making in a fuzzy environment. *Management Sci.* **17** (4), B-141–64, Dec. 1970.

Bibliography on Fuzzy Sets, their Applications, and Closely Related Topics

Adams, E. W. Elements of a theory of inexact measurement. *Phil. Sci.* 205–228, 1965.

Adavič, P. N., Borisov, A. N., and Golender, V. E. An adaptive algorithm for recognition of fuzzy patterns. *Kibernet. Diagnost.* 13–18, 1968 (in Russian).

Adey, W. R. Organization of Brain Tissue: Is the brain a noisy processor? *Int. J. Neurosci.* 271–284, 1972.

Albin, M. *Fuzzy Sets and their Application to Medical Diagnosis.* Ph. D. Thesis, University of California, Berkeley, Calif., 1975.

Allen, A. D. A method of evaluating technical journals on the basis of published comments through fuzzy implications: a survey of the major IEEE transactions. *IEEE Trans. Syst. Man Cybernet.* **SMC-3**, 422–5, 1973.

Allen, A. D. Measuring the empirical properties of sets. *IEEE Trans. Syst. Man Cybernet.* **SMC-4**, 66–73, 1974.

Arbib, M. A. *Semi-Ring Languages.* Standford University, Stanford, Calif, 1970.

Arbib, M. A., and Manes, E. G. Fuzzy morphisms in automata theory. *Proc. 1st Int. Symp. on Category Theory Applied to Computations and Control,* 98–105, 1974.

Arbib, M. A., and Manes, E. G. *Fuzzy Machines in a Category.* Coins Tech. Rep. No. 75B-1, University of Massachusetts, Amherst, Mass., 1975.

Arbib, M. A., and Manes, E. G. A category-theoretic approach to systems in a fuzzy world. *Synthese* 381–406, 1975.

Asai, K. and Kitajima, S. A method for optimizing control of multimodal systems using fuzzy automata. *Inform. Sci.* 343–53, 1971.

Asai, K., and Kitajima, S. Learning control of multimodal systems by automata. In *Pattern Recognition and Model Learning.* Plenum Press, New York, 1971.

Asai, K., and Kitajima, S. Optimizing control using fuzzy automata. *Automatica* 101–4, 1971.

Assillian, S. *Artificial Intelligence in the Control of Real Dynamic Systems.* Ph.D. Thesis, Queen Mary College, London, 1974.

Aubin, J. P. *Fuzzy Games*. University of Wisconsin, Madsion, Wis., 1974.

Aubin, J. P. Theorie de jeux. *C. R. Acad. Sci. Paris* **279**, A-891, 1974.

Aubin, J. P. Theorie de jeux. *C. R. Acad. Sci. Paris* **279**, A-963, 1974.

Baas, S. M., and Kwakernaak, H. *Rating and Ranking of Multiple-Aspect Alternatives using Fuzzy Sets*. Memo. No. 73, Dept. of Applied Mathematics, Twente University of Technology, Enschede, The Netherlands, 1975.

Banaschewska, B. Injective hulls in the category of distributive lattices. *J. Reine Angew. Math.* 102–9, 1968.

Banaschewski, B., and Bruns, G. Categorical characterization of the Mac-Neville Completion. *Arch. Math.* 369–77, 1967.

Bang, S. Y., and Yeh, R. T. *Toward a Theory of Relational Data Structure,* Rep. No. SELTR-1, University of Texas, Austin, Texas, 1974.

Barnev, P., Dimitrov, V., and Stanchev, V. *Fuzzy System Approach to Decision-Making based on Public Opinion Investigation through Questionnaires*. Institute of Mathematics and Mechanics, Bulgarian Academy of Sciences, Sofia, Bulgaria, 1974.

Becker, J. M. *A Structural Design Process*. Thesis, Dept. of Civil Engineering, University of California, Berkeley, Calif. 1973.

Bellmann, R. E. *Law and Mathematics*. Tech. Rep. No. 71-34, University of Southern California, Los Angeles, Calif., 1971.

Bellmann, R. E. *Local Logics,* Tech. Rep. No. 74-9, University of Southern California, Los Angeles, Calif., 1974.

Bellmann, R. E., and Giertz, M. On the analytic formalism of the theory of fuzzy sets. *Inform. Sci.* 149–56, 1973.

Bellmann, R. E., Kalaba, R., and Zadeh, L. A. Abstraction and pattern classification. *J. Math. Analysis Applic.* 1–7, 1966.

Bellmann, R. E., and Marchi, E. *Games of Protocol: The City as a Dynamic Competetive Process,* Tech. Rep. No. RB73-36, University of Southern California, Los Angeles, Calif., 1973.

Bellmann, R. E., and Zadeh, L. A. Decision-making in a fuzzy environment. *Management Sci.* B141–B-164, 1970.

Bezdek, J. C., *Fuzzy Mathematics in Pattern Classification*. Thesis, Center for Applied Mathematics, Cornell University, Ithaca, N.Y., 1973.

Bezdek, J. C. Numerical taxonomy with fuzzy sets. *J. Math. Biol.* **1**, 57–71, 1974.

Bezdek, J. C. Cluster validity with fuzzy sets. *J. Cybernet.* 58–73, 1974.

Bezdek, J. C. Mathematical Models for Systematics and Taxonomy. *Proc. 8th Annual Int. Conf. on Numerical Taxonomy, San Francisco,* 1975.

Bezdek, J. C., and Dunn, J. C. *Optimal Fuzzy Partitions: A Heuristic for Estimating the Parameters in a Mixture of Normal Distributions.* Cornell University, Ithaca, N.Y. 1974.

Black, M. "Vagueness": An exercise in logical analysis. *Phil. Sci.* B141–64, 1970.

Black, M. *Margins of Precision.* Cornell University Press, Ithaca, N.Y., 1970.

Black, M. Reasoning with loose concepts. *Dialogue* **2**, 1–12, 1973.

Blin, J. M., and Whinston, A. B. Fuzzy sets and social choice. *J. Cybernet.* **3**, 29-33, 1973.

Blin, J. M., and Whinston, A. B. Fuzzy sets and social choice. *J. Cybernet.* **3**, 29-33, 1974.

Boicescu, V. Sur les algebres de Lukasiewicz. *Logique, Automatique, Informatique,* 71–97. Academie Republique Socialiste de Roumanie, 1971.

Borghi, O. On a theory of functional probability, *Revta Un. Mat. Argentina* 90–106, 1972.

Borisov, A. N., and Kokle, E. A. Recognition of fuzzy patterns. *Kibernet. Diagnost.* 135–47, 1970.

Borisov, A. N., Osis, J. J. Methods for Experimental Estimation of Membership functions of fuzzy sets. *Kibernet. Diagnost.* 125–34, 1970.

Borisov, A. N., Vulf, G. N., and Osis, J. J. Prediction of the state of a complex system using the theory of fuzzy sets. *Kibernet. Diagnost.* 79–84, 1972.

Bossel, H. H., and Hughes, B. B. *Simulation of Value-Controlled Decision-Making.* Rep. No. SRC 73-11, Systems Research Center, Case Western Reserve University, Cleveland, Ohio, 1973.

Bremermann, H. J. Cybernetic Functionals and Fuzzy sets. *IEEE Symp. Record, Systems, Man and Cybernetics,* 248–53, 1971.

Bremermann, H. J. *Pattern Recognition.* Location and date unknown.

Brown, J. G. *Fuzzy Sets on Boolean Lattices.* Memo. Rep. No. 1957, Ballistic Research Laboratories, 1969.

Brown, J. G. A note on fuzzy sets. *Inform. Control* 32–9, 1971.

Bunge, M. C. *Categories of Sets Valued Functors,* Ph.D. Thesis, University of California, Berkeley, Calif., 1966.

Butnariu, D. L-fuzzy automata description of a neural model. *Proc. 3rd Int. Congr. of Cybernetics and Systems, Bucharest,* 1975.

Capocelli, R. M., and De Luca, A. Measures of uncertainty in the context of fuzzy sets theory. *Atti 2nd Congr. Nat. di Cibernetica die Casciana Terme, Pisa, Italy,* 1972.

Capocelli, R. M., and De Luca, A. Fuzzy sets and decision theory (abstr.). *Not. Am. Math. Soc.* A-709, 1972.

Capocelli, R. M., and De Luca, A. Fuzzy sets and decision theory. *Inform. Control,* 446–73, 1973.

Carlucci, D., and Donati, F. A fuzzy cluster of the demand within a regional service system. *Special Interest Discussion on Fuzzy Automata and Decision Processes, 6th IFAC World Congr., Boston, Mass.,* 1975.

Carlstrom, I. F. Truth and entailment for a vague quantifier. *Synthese,* 461–95.

Castonguay, C. *Meaning and Existence in Mathematics, Library of Exact Philosophy,* Vol. 9. Springer-Verlag, Berlin, 1972.

Chang, C. C. Infinite valued logic as a basis for set theory. *Proc. 1964 Int. Congr. for Logic, Methodology and Philosophy of Science.* North-Holland, Amsterdam, 1964.

Chang, C. L. *Fuzzy Sets and Pattern Recognition.* Thesis, University of California, Berkeley, Calif, 1967.

Chang, C. L. Fuzzy topological spaces. *J. Math. Analysis Appl.* 182–90, 1968.

Chang, C. L. *Fuzzy Algebras, Fuzzy Functions and their Applications to Function Approximation.* Dept. of Health, Education and Welfare, National Institute of Health, Bethesda, Md., 1970.

Chang, C. L. Interpretation and execution of fuzzy programs. In *Fuzzy Sets and their Applications to Cognitive and Decision Processes* (Ed. L. A. Zadeh *et al.*), 191–218. Academic Press, New York, 1975.

Chang, S. K. Picture processing grammar and its applications. *Inform. Sci.* 121–48, 1971.

Chang, S. K. Automated interpretation and editing of fuzzy line drawings. *Proc. Spain Joint Computer Conf.,* 393–9, 1971.

Chang, S. K. Fuzzy programs—theory and applications. *Proc. Symp. on Computers and Automata,* 147–61. Polytechnic Press of the Institute of Brooklyn, New York, 1971.

Chang, S. K. On the execution of fuzzy programs using finite-state machines. *IEEE Trans. Computers,* 241–53, 1972.

Chang, S. S. L. Fuzzy dynamic programming and the decision making process. *Proc. 3rd Princeton Conf. on Information Science and Systems,* 200–3. Princeton University Press, Princeton, N.J., 1969.

Chang, S. S. L. Fuzzy mathematics, man, and his environment. *IEEE Trans. Syst. Man Cybernet.* **SMC-2,** 92–3, 1972.

Chang, S. S. L. On fuzzy algorithm and mapping. *Special Interest Discussion on Fuzzy Automata and Decision Processes, 6th IFAC World Congr., Boston, Mass.,* 1975.

Chang, S. S. L. On risk and decision making in a fuzzy environment. In *Fuzzy Sets and their Applications to Cognitive and Decision Processes* (Ed. L. A. Zadeh *et al.*), 219–26. Academic Press, New York, 1975.

Chang, S. S. L., and Zadeh, L. A. On fuzzy mapping and control. *IEEE Trans. Syst. Man Cybernet.* **SMC-2,** 30–4, 1972.

Chapin, E. W. An axiomization of the set theory of Zadeh. *Not. Am. Math. Soc.* **687-02-4,** 753, 1971.

Chapin, E. W. Set-valued set theory: I. *Notre Dame J. Formal Logic* 255–67, 1974.

Chapin, E. W. Set-valued set theory: II. *Notre Dame J. Formal Logic* 244–67, 1975.

Chen, C. Realizability of Communications nets: An application of the Zadeh criterion. *IEEE Trans. Circuits Syst.* **CAS-21,** 150–1, 1974.

Cignoli, R. *Estudio Algebraice de Logicas Polivalentes: Algebras de Moisil de Orden N.* Ph.D. Thesis, Universidad Nacionale del Sur-Bahia-Blanca, Argentina.

Cleave, J. P. The notion of logical consequence in the logic of inexact predicates. *Z. Math. Logik. Grumd. Math.* 307–24, 1974.

Cleave, J. P. The notion of validity in logical systems with inexact predicates. *Br. J. Phil. Sci.* 269–74, 1970.

Cohen, P. J., and Hirsch, R. Non-Cantorian set theory. *Sci. Am.* 101–16, 1967.

Conche, B. Elements d'une methode de classification par utilisation d'un automate flou. *JEEFLN,* Université de Paris-Dauphine, Paris, 1973.

Conche, B., Jouault, J. P., and Luan, P. M. Application des concepts flous a la programmation en languages quasi-naturels. *Seminaire Bernard Roy.* Université de Paris-Dauphine, Paris, 1973.

Cools, M. La semantique dans les processus didactiques. *Imago Disc. Pap.* Université Catholique de Louvain, Belgium, 1973.

Cools, M., and Peteau, M. STIM 5: Un programme de stimulation inventive utilisant la theorie des sous-ensembles flous. *Imago Disc. Pap.* Université Catholique de Louvain, Belgium, 1973.

Dal Cin, N. *Fuzzy-State Automata, their Stability and Fault-Tolerance.* Institute of Science, University of Tubingen, Germany, 1975.

Dal Cin, N. Modification tolerances of fuzzy-state automata. *Int. J. Computer Inform. Sci.* 81–93, 1975.

Davio, M., and Thayse, A. Representation of fuzzy functions. *Philips Res. Rep.* 93–106, 1973.

De Kerf, J. Vage verzamelingen. *Omega (Veren. Wis- Natuurkund. Lovan.)* 2–18, 1974.

De Kerf, J. Vage verzamelingen. *Ingenieurstijdingen* 581–89, 1974.

De Luca, A., and Termini, S. Algorithmic aspects in complex systems analysis. *Scientia* **106,** 659–71, 1971.

De Luca, A., and Termini, S. Algebraic properties of fuzzy sets *J. Math. Analysis Applic.* 373–86, 1972.

De Luca, A., and Termini, S. A definition of a nonprobabilistic entropy in the setting of fuzzy set theory. *Inform. Control* 301–312, 1972.

De Luca, A., and Termini, S. Entropy of L-fuzzy sets. *Inform. Control* 55–73, 1974.

De Palma, G. F., and Yau, S. S. Fractionally fuzzy grammars with application to pattern recognition. In *Fuzzy sets and their Applications to Cognitive and Decision Processes* (Ed. L. a. Zadeh *et al.*), 329–50, Academic Press, New York, 1975.

Diamond, P. *Fuzzy Chaos.* Dept. of Mathematics, University of Queensland, Brisbane, Australia.

Diday, E. Optimisation en classification automatique et reconnaissance des formes. *Not. Sci.* **IRIA-6,** Bull. 12.

Dimitrov, V. *Efficient Governing Heuristic Systems by Fuzzy Instructions.* Dept. of Software, Institute of Mathematics, Bulgarian Academy of Sciences, Sofia, Bulgaria, 1975.

Dimitrov, V., Wechler, W., and Barnev, P. *Optimized Fuzzy Control of Humanistic Systems.* Dept. of Mathematics, Dresden Technical University, Dresden, German Democratic Republic, 1974.

Dimitrov, V., Wechler, W., and Barnev, P. *Optimal Fuzzy Control of Humanistic Systems*. Institute of Mathematics and Mechanics, Bulgarian Academy of Sciences, Sofia, Bulgaria, 1974.

Dravecky, J., and Riecan, B. Measurability of functions with values in partially ordered spaces, *Cas. Pestovani Math.* 27–359, 1975.

Drosselmeyer, E., and Wonneberger. Studies on a fuzzy system in the parochial field: *Special Interest Discussion on Fuzzy Automata and Decision Processes, 6th IFAC World Congr. Boston, Mass. 1975.*

Dubois, T. *Une Methode d'evaluation par les sous-ensembles flous appliquée à la simulation. Imago Disc. Pap. 13,* Université Catholique de Louvain, Louvain, Belgium, 1974.

Dunn, J. C. A graph theoretic analysis of pattern classification via Tamura's fuzzy relation. *IEEE Trans. Syst. Man Cybernet.* **SMC-4,** 310–3, 1974.

Dunn, J. C. *A Fuzzy Relative of the Isodata Process and its Use in Detecting Compact Well-Separated Clusters.* Cornell University, Ithaca, N.Y.

Dunn, J. C. *Some Recent Investigations of a New Fuzzy Partitioning Algorithm and its Application to Pattern Classification Problems.* Cornell University, Ithaca, N.Y., 1975.

Dunn, J. C. *Well Separated Clusters and Optimal Fuzzy Partitions.* Cornell University, Ithaca, N.Y.

Dunn, J. C. *Canonical Forms of Tamura's Fuzzy Relation Matrix: A Scheme for Visualizing Cluster Hierarchies.* Cornell University, Ithaca, N.Y., 1975.

Dunn, J. C. Indices of partition fuzziness and the detection of clusters in large data sets: *Special Interest Discussion on Fuzzy Automata and Decision Processes, 6th IFAC World Cong., Boston, Mass.,* 1975.

Dunst, A. J. *Application of the Fuzzy Set Theory.* Location Unknown, 1971.

Endo, Y., and Tsukamoto, Y. Apportion models of tourists by fuzzy integrals. *Annual Conf. Rec. SICE,* 1973.

Engel, A. B., and Buonamano, V. *Towards a General Theory of Fuzzy Sets. I.* Université Estadual de Campinas, São Paulo, Brazil, 1973.

Engel, A. B., and Buonamano, V. *Towards a General Theory of Fuzzy Sets. II.* Université Estadual de Campinas, São Paulo, Brazil.

Esogbue, A. M. O. *Dynamic Programming and Fuzzy Allocation Processes.* Tech. Memo No. 202, Operational Research Dept., Case Western Reserve University, Cleveland, Ohio, 1970.

Esogbue, A. O. *On Application of Fuzzy Allocation Theory to the Modelling of Cancer Research Appropriation Process.* Georgia Institute of Technology, Atlanta, Ga., 1975.

Fellinger, W. L. *Specifications for a Fuzzy System Modelling Language.* Thesis, Oregon State University, Corvallis, Ore., 1974.

Fine, K. Vagueness, truth and logic. *Synthese* 265–300, 1975.

Floyd, R. W. Non-deterministic algorithms. *JACM* 636–44, 1967.

Fu, K. S. A critical review of learning control research. In *Pattern Recognition and Machine Learning.* Plenum Press, New York, 1971.

Fu, K. S. *Pattern Recognition and some Socio-Economic Problem.* Purdue University, West Lafayette, Ind, 1974.

Fu, K. S., and Li, T. J. Formulation of learning automata and games. *Information Sci.* 237–56, 1969.

Fujisake, H. Fuzziness in medical sciences and its processing. *Proc. Symp. on Fuzziness in Systems and its Processing (SICE),* 1971.

Fung, L. W., and Fu, K. S. The *K*th optimal policy algorithm for decision making in fuzzy environments. In *Identification and System Parameter Estimation,* 1052–9, North-Holland, Amsterdam, 1974.

Fung, L. W., and Fu, K. S. Characterisation of a class of fuzzy optimal control problems. *Proc. 8th Princeton Conf. on Information Science and Systems.* Princeton University Press, Princeton, N.J., 1974.

Fung, L. W., and Fu, K. S. An axiomatic approach to rational decision making. In *Fuzzy Sets and their Applications to Cognitive and Decision Processes* (Ed. L. A. Zadeh *et al.*), 227–56. Academic Press, New York, 1975.

Fung, L. W., and Fu, K. S. *Decision-Making in a Fuzzy Environment.* Purdue University, West Lafayette, Ind.

Furukawa, M., Nakamura, K., and Oda, M. Fuzzy model of human decision-making process. *Annual Conf. Records of SICE,* 1972.

Furukawa, M., Nakamura, K., and Oda, M. Fuzzy variant process of memories. *Annual Conf. Rec. SICE,* 1973.

Gaines, B. R. Stochastic and fuzzy logics. *Electron. Lett.* 188–9, 1975.

Gaines, B. R. *Behaviour/Structure Transformations under Uncertainty.* Rep. No. EES-MMS-AUT-75, Dept. of Electrical Engineering Science, University of Essex, U.K., 1975.

Gaines, B. R. *Fuzzy and Stochastic Metric Logics*. Rep. No. EFS-MMS-FU71-75, Dept. of Electrical Engineering Science, University of Essex, U.K.

Gaines, B. R. Fuzzy reasoning and the logic of uncertainty. *Proc. 6th Int. Symp. on Multiple-Valued Logic, Logan, Utah*, 1976.

Gaines, B. R. *A Bibliography of Fuzzy Systems Theory and Closely Related Topics*. Dept. of Electrical Engineering Science, University of Essex, U.K., 1976.

Gaines, B. R. *Multivalued Logics and Fuzzy Reasoning: Workshop on Discrete Systems and Fuzzy Reasoning*. Dept. of Electrical Engineering Science, University of Essex, U.K., 1976.

Gaines, B. R. *Survey of Fuzzy Reasoning: Workshop on Discrete Systems and Reasoning,* Electrical Engineering, Queen Mary College, London, 1976.

Gearing, Ch. E. *Generalized Bayesian Posterior Analysis with Ambiguous Information*. Dartmouth College, Hanover, N.H.

Gentilhomme, Y. Les ensembles flous en linguistique. *Cah. Linguist. Theorie Appl.* Bucharest, 1968.

Georgescu, G. Les algebres de Lukasiewicz 0-valents. *Logique, Automatique, Informatique,* 99–176. Academic Republique Socialiste de Roumanie, 1971.

Gerhardts, M. D. Zur charakterisierung distributiver scheirverdande. *Math. Annln.* 231–46, 1965.

Gerhardts, M. D. Schragverdande und quasiordnungen. *Math. Annln.* 65–73, 1969.

Gluss, B. Fuzzy multistage decision making and terminal regulators and their relationship to non-fuzzy quadratic state and terminal regulators. *Int. J. Control* 177–92, 1973.

Giles, F. A logic for subjective belief. *Proc. Conf. on Foundation of Probability and Statistics and Statistical Theories of Science, London, Ontario, 1973.*

Giles, R. *Lukasiewicz's Logic and Fuzzy Set Theory*. Preprint 1974–29, Dept. of Mathematics, Queen's University, Kingston, Canada, 1974.

Giles, R. *A Nonclassical Logic for Physics,* Studia Logica, 1974.

Gitman, I. *Organization of Data: A Model and Computational Algorithm that uses the Notation of Fuzzy Sets*. Ph.D. Thesis, McGill University, Montreal, Canada, 1970.

Gitman, I., and Levine, M. D. An algorithm for detecting unimodal fuzzy sets and an application as a clustering technique, *IEEE Trans. Computers* 583–93, 1970.

Gluss, B. Fuzzy multistage decision-making; fuzzy states and terminal regulators and their relationship to non-fuzzy quadratic stage and terminal regulators. *Int. J. Control* 177–92, 1973.

Goguen, J. A. L-fuzzy sets. *J. Math. Analysis Applic.* 145–74, 1967.

Goguen, J. A. *Categories of V-sets*. University of California, Berkeley, Calif., 1968.

Goguen, J. A. *Categories of Fuzzy Sets: Applications of Non-Cantorian Set Theory*. Ph.D. Thesis, University of California, Berkeley, Calif., 1968.

Goguen, J. A. The logic of inexact concepts. *Synthese,* 326–73, 1968.

Goguen, J. A. *Representing Inexact Concepts*. ICR Quarterly Rep. No. 20, Institute for Computer Research, University of Chicago, Chicago, Ill. 1969.

Goguen, J. A. Mathematical representation of hierarchically organized systems. *Global Systems Dynamics (Ed. E. O. Attinger), 111–29.*

Goguen, J. A. Hierarchical inexact data structures in artificial intelligence problems. *Proc. 5th Hawaii Int. Conf. on System Sciences, Honolulu, Hawaii,* 345–7, 1972.

Goguen, J. A. *Some Comments on Applying Mathematical System Theory.* Doc. Computer Science Dept., University of California, Los Angeles, Calif., 1973.

Goguen, J. A. Systems theory concepts in computer science. *Proc. 6th Hawaii Int. Conf. on Systems Sciences, Honolulu, Hawaii,* 77–80, 1973.

Goguen, J. A. *Axioms, Extensions and Applications for Fuzzy Sets: Languages and Representation on Concepts*. Doc. Computer Sciences Dept., University of California, Los Angeles, Calif., 1973.

Goguen, J. A. The fuzzy Tychonoff theorem. *J. Math. Analysis Applic.,* 734–42, 1973.

Goguen, J. A. Concept representation in natural and artificial languages: Axioms, extensions and applications for fuzzy sets. *Int. J. Man–Machine Studies* 513–61, 1974.

Goguen, J. A. On fuzzy robot planning. In *Fuzzy Sets and their Applications to Cognitive and Decision Processes* (Ed. L. A. Zadeh *et al.*), 429–48. Academic Press, New York, 1975.

Goguen, J. A. Complexity of Hierarchically organized systems and the structure of musical experiences. *UCLA Q. Rep. Dept. Computer Sci.,* 1975.

Goguen, J. A., Thatcher, J., Wagner, E., and Wright, J. A junction between computer science and category theory: basic concepts and examples: II. Universal constructions. *IBM Res. Rep.,* 1973.

Goodman, J. S. *From Multiple Balyage to Fuzzy Sets.* Institute of mathematics, University of Florence, Italy.

Gottinger, H. W. A fuzzy algorithmic approach to the definition of complex or imprecise concepts. *Conf. on Systems Theory, University of Bielefeld,* 1975.

Gottinger, H. W. *Toward a Fuzzy Reasoning in the Behavioral Science,* University of California, Santa Barbara, Calif.

Gottwald, S. Zahlbereighskonstruktionen in einer mehrwertigen Mengenlehre. *Z. Math. Logik Grumd. Math.* 145-88, 1971.

Gottwald, S. Fuzzy topology: Product and quotient theorems. *J. Math. Analysis Applic.* 512-21, 1974.

Grattan-Guiness *Fuzzy Membership Mapped onto Interval and Many-Valued Quantities.* Unpublished.

Gusev, L. A., and Smirnova, I. M. Fuzzy sets: Theory and applications (A survey). *Automat. Telemekh.* 66-85, 1973 (in Russian).

Gusev, L. A., and Smirnova, J. M. Simulation of behavior and intelligence. *Automat. Telemekh.* 66-85, 1973 (English translation).

Hamacher, H. *Uber Logische Verknupfungen Unscharfer Aussagen und deren Zugehorige Bewertungsfunktionen.* Rep. No. 75/14, Lehrstuhl fur Unternehmensforschung, RWTH Aachen, German Federal Republic, 1975.

Harris, J. I. *Fuzzy Implication Comments on a Concept of Zadeh,* DOAE Research Working Paper CM, Ministry of Defence, Byfleet, Surrey, U.K.

Harris, J. I. *Fuzzy Sets: How to be Imprecise Precisely.* DOAE Research Working Paper CM5, Ministry of Defence, Byfleet, Surrey, U.K.

Hendry, W. L. *Fuzzy Sets and Russell's Paradox.* Los Alamos Science Laboratories, University of California, Los Alamos, N.M.

Henry-Labordere, A., de Backer, E. *Intelligence Creatrice: Automatisation de Processus d'Association.* Revue metra int. Rep. No. 69.

Hirai, H., Asai, K., and Katajima, S. Fuzzy automaton and its application to learning control systems. *Mem. Fac. Eng. Osaka City Univ.* 67-73, 1968.

Honda, N. Fuzzy sets, *J. IECE,* 1359–63, 1971.

Honda, N., and Nasu, M. Recognition of fuzzy languages. In *Fuzzy Sets and their Applications of Cognitive and Decision Processes* (Ed. L. A. Zadeh *et al.*), 279–300. Academic Press, New York, 1975.

Honda, N., and Nasu, M. F-recognition of fuzzy languages. *Special Interest Discussion on Fuzzy Automata and Decision Processes, 6th IFAC World Congr.* Boston, Mass., 1975.

Horejŝ, J. Classifications and their relationship to a measure. *Publ. Faculty Sci. Univ. Purkyne (Brno)* No. 168, 475–93, 1965.

Hormann, A. M. *Machine-Aided Value Judgments using Fuzzy Set Techniques.* Rep. No. SP-3590, System Development Corporation, Santa Monica, Calif, 1971.

Hughes, G. E., and Creswell, M. J. *An Introduction to Modal Logic.* Methuen, London, 1968.

Hughes, J. S., and Kandel, A. Applications of Fuzzy Algebra to Hazard Detection in Combinatorial Switching Circuits, Rep. No. CSR 138, Dept. of Computer Science, New Mexico Institute of Mining and Technology, Socorro, N.M., 1975.

Hung, N. T. *Information Fontionelle et Ensembles Flous, Seminar on Questionnaires.* University of Paris, Paris, 1975.

Hutton, B. *Normality in Fuzzy Topological Spaces,* Dept. of Mathematics, University of Auckland, New Zealand, 1974.

Hutton, B., and Reilly, J. L. *Separation Axioms in Fuzzy Topological Spaces.* University of Auckland, New Zealand, 1974.

Inagaki, Y., and Fukumura, T. On the description of fuzzy meaning of context-free language. In *Fuzzy Sets and their Applications to Cognitive and Decision Processes* (Ed. L. A. Zadeh *et al.*), 301–28. Academic Press, New York, 1975.

Ishikawa, A., and Mirno, H. Design of a video information system and the fuzzy information theory. *Eurocomp 75, Brunel University, 1975.*

Jacobson, D. H. *On Fuzzy Goals and Maximizing Decisions in Stochastic Optimal Control.* CSID Special Rep., National Research Institute for Mathematical Science, Pretoria, South Africa, 1975.

Jain, R. Outline of an approach for the analysis of fuzzy systems. *Special Interest Discussion on Fuzzy Automata and Decision Processes, 6th IFAC World Congr., Boston, Mass.,* 1975.

Jain, R. Convolution of fuzzy variables, *J. IETE,* **22,** 1976.

Jain, R. Decision making with fuzzy knowledge about the state of the system. *Nation System Conf., University of Roorkee, India,* 1976.

Jakubowski, R., and Kasprzak, A. Application of fuzzy programs to the design of machining technology. *Bull. Polish Acad. Sci.* 17–22, 1973.

Jarvis, P. A. Optimization strategies in adaptive control: a selective survey. *IEEE Trans. Syst. Man Cybernet.* **SMC-5,** 83–94, 1975.

Jouault, J. P., and Pham Minh Luan. *Application des Concepts Flous à la Programmation en Langages Quasi-Naturels.* Institute d'Informatique d'Enterprise, CNAM, Paris, France, 1973.

Kahne, S. A procedure for optimizing development decisions. *Automatica* 261–9, 1975.

Kalmanson, D. *Recherche Cardio-Vasculaire et Theorie des Ensembles Flous,* 2757–60. Nouvelle Presse Medicale, Paris, 1973.

Kandel, A. *Toward Simplification of Fuzzy Functions.* Rep. No. CSR 114, New Mexico Institute of Mining and Technology, Socorro, N.M., 1972.

Kandel, A. *On Coded Grammars and Fuzzy Structures.* Rep. No. CSR 118, New Mexico Institute of Mining and Technology, Socorro, N.M., 1972.

Kandel, A. *Fuzzy Chains: A New Concept in Decision-Making under Uncertainty,* Rep. No. CSR 123, New Mexico Institute of Mining and Technology, Socorro, N.M., 1973.

Kandel, A. *A New Method for Generating Fuzzy Prime Implicants and an Algorithm for the Automatic Minimization of Inexact Structures.* Rep. No. CSR 126, Dept. of Computer Science, New Mexico Institute of Mining and Technology, Socorro, N.M., 1973.

Kandel, A. *A New Algorithm for Minimizing Incompletely Specified Fuzzy Functions.* Rep. No. CSR 127, New Mexico Institute of Mining and Technology, Socorro, N.M., 1973.

Kandel, A. On the analysis of fuzzy logic. *Proc. 6th Int. Conf. on System Sciences, Honolulu, Hawaii,* 1973.

Kandel. A. Comment on an algorithm that generates fuzzy prime implicants by Lee and Chang. *Inform. Control* 279–82, 1973.

Kandel, A. Fuzzy Functions and their application to the analysis of switching hazards. *Proc. 2nd Texas Conf. on Computing Systems, Austin, Texas,* 42/1–42/6, 1973.

Kandel, A. On minimization of fuzzy functions. *IEEE Trans. Computers,* **C-22,** 826–32, 1973.

Kandel, A. Comments on"Minimization of fuzzy functions." *IEEE Trans. Computers* **C-22,** 217, 1973.

Kandel, A. *On Fuzzy Maps: Some Initial Thoughts.* Rep. No. CSR 131, Dept. of Computer Science, New Mexico Institute of Mining and Technology, Socorro, N.M., 1974.

Kandel, A. *Simple Disjunctive Decomposition of Fuzzy Functions.* Rep. No. CSR 132, Dept. of Computer Science, New Mexico Institute of Mining and Technology, Socorro, N.M., 1974.

Kandel, A. *On the Theory of Fuzzy Matrices.* Rep. No. CSR 135, Dept. of Computer Science, New Mexico Institute of Mining and Technology, Socorro, N.M., 1974.

Kandel, A. *Generation of the Set Representing all Fuzzy Prime Implicants.* Rep. No. CSR 136, Dept. of Computer Science, New Mexico Institute of Mining and Technology, Socorro, N.M., 1974.

Kandel, A. On the Minimization of incompletely specified fuzzy functions. *Inform. Control* 141–53, 1974.

Kandel, A. Codes over languages. *IEEE Trans. Man Cybernet.* **SMC-4,** 125–38, 1974.

Kandel, A. Fuzzy representation, CNF minimization, and their application to fuzzy transmission structures. *Proc. 1974 Int. Symp. on Multiple-Valued Logic, Morgantown, W. Va.,* 361–79, 1974.

Kandel, A. Synthesis of fuzzy logic with analog modules: preliminary developments. *Computers Educ. Trans. (ASEE)* 71–9, 1974.

Kandel, A. Application of fuzzy logic to the detection of static hazards in combinational switching systems. *Int. J. Computer Inform. Sci.* 129–39, 1974.

Kandel, A. On the Enumeration of fuzzy functions. *12th Holiday Symp. on Developments in Combinatorics, New Mexico State University,* 1974.

Kandel, A. On the properties of fuzzy switching functions, *J. Cybernet.* 119–26, 1974.

Kandel, A. *A Note on the Simplification of Fuzzy Switching Functions.* Rep. No. CSR 139, Dept. of Computer Science, New Mexico Institute of Mining and Technology, Socorro, N.M., 1975.

Kandel, A. Fuzzy hierarchical classification of dynamic patterns. *NATO A.S.I. on Pattern Recognition and Classification, France,* 1975.

Kandel, A. Block decomposition of imprecise models. *Conf. Rec. 9th Asilomar Conf. on Circuits, Systems, and Computers, Pacific Grove, Calif.*, 1975.

Kandel, A. Properties of fuzzy matrices and their applications to heirarchical structures. *Conf. Rec. 9th Asilomar Conf. on Circuits, Systems, and Computers, Pacific Grove, Calif.*, 1975.

Kandel, A. Fuzzy systems and their applications to simulations. *Proc. 9th Hawaii Int. Conf. on System Sciences, Honolulu, Hawaii*, 1976.

Kandel, A. Inexact switching logic. *IEEE Trans. Syst. Man Cybernet.* **SMC-6,** 215–9, 1976.

Kandel, A. On the decomposition of fuzzy functions. *IEEE Trans. Computers,* **C-25,** 1124-30, 1976.

Kandel, A. Fuzzy maps and their applications in the simplification of fuzzy switching functions. *Proc. 6th Int. Symp. on Multiple-Valued Logic, Logan, Utah,* 1976.

Kandel, A., and Davis, H. A. The first fuzzy decade (bibliography on fuzzy sets and their applications). Rep. No. CSR 140, Dept. of Computer Science, New Mexico Institute of Mining and Technology, Socorro, N.M., 1976.

Kandel, A., and Obenauf, T.A. *On Fuzzy Lattices.* Rep. No. CSR 128, New Mexico Institute of Mining and Technology, Socorro, N.M., 1973.

Kandel, A. and Yelowitz, L. Fuzzy chains. *IEEE Trans. Systems Man Cybernet.* 472–5, 1974.

Kaufmann, A. *Introduction à la Theorie des Sous-Ensembles Flous (à l'Usage des Ingenieurs).* Vol. 1. Elements Theoriques de Base. Masson, Paris, 1973. (English Transl. published by Academic Press, New York, 1975).

Kaufmann, A. *Introduction à la Theorie des Sous-Ensembles Flous (à l'Usage des Ingenieurs).* Vol. 2. *Applications à la Linguistique, à la Logique et à la Semantique.* Masson, Paris, 1975.

Kaufmann, A. *Introduction à la Theorie des Sous-Ensembles Flous (à l'Usage des Ingenieurs).* Vol. 3. *Applications à la Classification et la Reconnaissance des Formes, aux Automates et aux Systemes, aux Choix des Critares.* Masson, Paris, 1975.

Kaufmann, A. Introduction to a fuzzy theory of the human operator. *Special Interest Discussion on Fuzzy Automata and Decision Processes, 6th IFAC World Congr., Boston, Mass.,* 1975.

Kaufmann, A. A survey of fuzzy sets theory and applications to languages, automata and algorithms. *US–Japan Seminar on Fuzzy Sets and their Applications, Berkeley, Calif.,* 1975.

Kaufmann, A., Cools, M., and Dubois, T. Stimulation inventive dans un dialogue homme–machine utilisant la methode des morphologies et la theorie des sous-ensembles flous. *Imago Disc. Pap.* IDP-6. Université Catholique de Louvain, Belgium, 1973.

Kaufmann, A., Cools, M., and Dubois, T. *Exercices avec Solutions sur la Theorie des Sous-Ensembles Flous.* Masson, Paris, 1975.

Khatchadourian, H. Vagueness, meaning and absurdity. *Am. Phil. Q.,* **2,** 119–29, 1965.

Kickert, W. J. M., and Van Nauta Lemke, H. R. *Application of a Fuzzy Controller in a Warm Water Plant.* Dept. of Electrical Engineering, Delft University of Technology, Holland, 1975.

Killing, E. *Fuzzy Planner.* Tech. Rep. No. 168, Computer Science Dept., University of Wisconsin, Madison, Wisc., 1973.

Kim, H. H., Mizumoto, M., Toyoda, J., and Tanaka, K. Lattice grammars. *Trans. IECE* 57-D, 1974.

King, P. J., and Mamdani, E. H. The application of fuzzy control systems to industrial processes. *Special Interest Discussion on Fuzzy Automata and Decision Processes, 6th IFAC World Congr., Boston, Mass.,* 1975.

Kise, V. A., and Osis, J. J. Search methods for establishing of maximal separability of fuzzy sets. *Kibernet. Diagnost.* 79–88, 1969.

Kitagawa, T. Three coordinate systems for information science approaches. *Inform. Sci.* 159–69, 1973.

Kitagawa, T. Biorobots for simulation studies of learning and intelligent controls. *US–Japan Seminar on Learning Control and Intelligent Control, Gainesville, Fla.,* 1973.

Kitagawa, T. Fuzziness in informative logics. In *Fuzzy Sets and their Applications to Cognitive and Decision Processes* (Ed. L. A. Zadeh *et al.*), 97–124. Academic Press, New York, 1975.

Kitajima, S., and Asai, K. Learning model of fuzzy automation with state-dependent output. *Annual Joint Conf. Rec. JAACE,* 1972.

Kitajima, S., and Asai, K. Learning controls by fuzzy automata. *J. JAACE* 551–9, 1970.

Kitajima, S., and Asai, K. A method of learning control varying search domain by fuzzy automata. *US-Japan Seminar on Learning Control and Intelligent Control, Gainesville, Fla.,* 1973.

Klaua, D. Uber einen Ansatz zur mehrwertigen Mengenlehre. *Monatsb. Dt. Akad. Wiss. Berl.* **7,** 859-67, 1965.

Klaua, D. Uber einen zweiten Ansatz zur mehrwertigen Mengenlehre. *Monatsb. Dt. Akad. Wiss. Berl.* **8,** 1966.

Klaua, D. Grundbegriffe einer mehrwertigen Mengenlehre. *Monatsb. Dt. Akad. Wiss. Berl.* **8,** 782-802, 1966.

Klaua, D. Ein Ansatz zur mehrwertigen Mengenlehre. *Math. Nachr.* **33,** 273-96, 1967.

Klaua, D. Einbettung der klassischen Mengenlehre in die Mehrwertige. *Monatsb. Dt. Akad. Wiss. Berl.* **9,** 258-72, 1967.

Klaua, D. Stetige Gleichmachtigkeiten Kontinuierlichwertiger Mengen. *Monatsb. Dt. Akad. Wiss. Berl.* **12,** 749-58, 1970.

Kling, R. *Fuzzy Planner,* Tech. Rep. No. 168, Computer Science Dept., University of Wisconsin, Madison, Wis., 1973.

Kling, R. Fuzzy planner: reasoning with inexact concepts in a procedural problem-solving language. *J. Cybernet.* 1-18, 1973.

Kling, R. Fuzzy planner: Computational inexactness in a procedural problem-solving language. Tech. Rep. No. 168, Dept.of Computer Science, University of Wisconsin, Madison, Wis., 1973.

Klir, G. J. On universal logic primitives. *IEEE Trans. Computers* **C-20,** 467-9, 1971.

Klir, G. J. *Processing of Fuzzy Activities of Neutral Systems.* State University of New York at Binghamton, Binghamton, N.Y.

Knopfmacher, J. On measures of fuzziness. *J. Math. Analysis Applic.* 529-34, 1975.

Kochen, M. On the precision of adjectives which denote fuzzy sets. Mental Health Research Institute, University of Michigan, Ann Arbor, Mich., 1974.

Kochen, M. Applications of fuzzy sets in psychology. In *Fuzzy Sets and their Applications to Cognitive and Decision Processes* (Ed. L. A. Zadeh *et al.*), 395-408, Academic Press, New York, 1975.

Kochen, M., and Dardre, A. N. On the precision of adjectives which denote fuzzy sets. *J. Cybernet.* 49-59, 1974.

Kochen, M., and Dreyfuss-Rami, G. *On the Psycholinguistic Reality of Fuzzy Sets: Effect of Context and Set*. Mental Health Research Institute, University of Michigan, Ann Arbor, Mich.

Koczy, L. T. *R-Fuzzy Algebra as a Generalized Formulation of the Intuitive Logic*. Dept. of Process Control, Technical University, Budapest, Hungary, 1975.

Koczy, L. T., and Hajnal, M. A new fuzzy calculus and its application as a pattern recognition technique. *Proc. 3rd Int. Congr. on Cybernetics and Systems, Bucharest*, 1975.

Kokawa, M., Nakamura, K., and Oda, M. Hint effect and a jump of logic in a decision process. *Trans. IECE Japan*.

Kokawa, M., Nakamura, K., and Oda, M. *Fuzzy Expression of Human Experience-to-Memory Process*. Res. Rep. Automatic Control Lab., Nagoya University, Nagoya, Japan, 1973.

Kokawa, M., Nakamura, K., and Oda, M. *Fuzzy-Theoretical Approaches to Forgetting Processes on Inference*. Nagoya University, Nagoya, Japan.

Kokawa, M., Nakamura, K., and Oda, M. Fuzzy Theoretical and concept formational approach to memory and inference experiments. *Trans. IECE Japan* **57-D,** 487–93, 1974.

Kokawa, M., Nakamura, K., and Oda, M. Experimental approach to fuzzy simulation of memorizing, forgetting and inference process. In *Fuzzy Sets and their Applications to Cognitive and Decision Processes* (Ed. L. A. Zadeh *et al.*), 409–28. Academic Press, New York, 1975.

Kokawa, M., Oda, M., and Nakamura, K. Fuzzy theoretical dimensionality reduction method of multi-dimensional quantity. *6th Triennial IFAC World Congr., Boston, Mass.,* 1975.

Kotoh, K., and Hiramatsu, K. A representation of pattern classes using the fuzzy sets. *Trans. IECE Japan* 275–82, 1973.

Lake, J. *Sets, Fuzzy Sets, Multi-Sets and Functions*. Dept. of Mathematics, Polytechnic of the South Bank, London, 1974.

Lake, J. *Fuzzy Sets and Bald Men.* Dept. of Mathematics, Polytechnic of the South Bank, London, 1974.

Lakoff, G. Hedges: a study in meaning criteria and the logic of fuzzy concepts. *J. Phil. Logic* 458–508.

Lakshmivarahan, S., and Rajasethupathy, K. *Considerations for Fuzzifying Formal Languages and Synthesis of Fuzzy Grammars.* Indian Institute of Technology, Madras, India, 1977.

Lee, E. T. *Fuzzy Languages and their Relation to Automata*. Thesis, Dept. of Electrical Engineering and Computer Science, University of California, Berkeley, Calif., 1972.

Lee, E. T. Proximity measures for the classification of geometric figures. *J. Cybernet.* 43–59, 1972.

Lee, E. T. The shape-oriented dissimilarity of polygons and its application to the classification of chromosome images. *Pattern Recog.* 47–60, 1974.

Lee, E. T. An application of fuzzy sets to the classification of geometric figures and chromosome images. *US–Japan Seminar on Fuzzy Sets and their Applications, Berkeley, Calif.,* 1974.

Lee, E. T. Shape-oriented chromosome classification. *IEEE Trans. Syst. Man Cybernet.* **SMC-5,** 629–32, 1975.

Lee, E. T. Shape oriented storage and retrieval of geometric figures and chromosome images. *Inform. Process. Management,* 1976.

Lee, E. T., and Zadeh, L. A. Note on the fuzzy languages. *Inform. Sci.* 421–34, 1969.

Lee, E. T., and Zadeh, L. A. Fuzzy languages and their acceptance by automata, *Proc. 4th Princeton Conf. on Information Science and Systems,* 399. Princeton University Press, Princeton, N.J., 1970.

Lee, R. T. C. Fuzzy logic and the resolution principle. *J. ACM* 109–19, 1972.

Lee, R. T. C, and Chang, C. L. Some properties of fuzzy logic. *Inform. Control* 417–31, 1971.

Lee, S. C. Fuzzy sets and neural networks. *US–Japan Seminar on Fuzzy Sets and their Application, Berkeley, Calif.,* 1974.

Lee, S. C., and Lee, E. T. Fuzzy neurons and automata. *Proc. 4th Princeton Conf. on Information Science and Systems,* 381–5, 1970.

Leenders, J. H. Vage verzamelingen: een kritishe benadering. *Kwart. Wetenschapp. Ond. Limburg* 441–55, 1974..

Lefaivre, R. The representation of fuzzy knowledge, *US–Japan Seminar on Fuzzy Sets and their Application, Berkeley, Calif., 1974.*

Le Faivre, R. *FUZZY: A Programming Language for Fuzzy Problem Solving.* Tech. Rep. No. 202, Dept. of Computer Science, University of Wisconsin, Madison, Wis., 1974.

Le Faivre, R. *Fuzzy Problem Solving.* Tech. Rep. No. 37, Madison Academic Computer Center, University of Wisconsin, Madison, Wis., 1974.

Lewis, B. Ordinary language algorithms and copying with fuzziness in every-day life. *Workshop on Discrete Systems and Fuzzy Reasoning*. Department of Electrical Engineering. Queen Mary College, London, 1976.

Lientz, B. P. On time dependent fuzzy sets. *Inform. Sci.* 367–76, 1972.

Loginov, V. I. Probability treatment of Zadeh membership functions and their use in pattern recognition. *Eng. Cybernet.* 68–9, 1966.

Lombaerde, J. Mesures d'entropie en theorie des sous-ensembles flous. *Imago Disc. Pap.* IDP-12. Centre Interfacultaire Imago, Université Catholique de Louvain, Belgium, 1974.

Longo, G. Fuzzy sets, graphs and source coding. In *New Directions in Signal Processing in Communications and Control* (Ed. J. K. Swirzynski), 27–33. Noordhoff, Leyden, 1975..

Lowen, M. R. *A Theory of Fuzzy Topologies*. Thesis, Dept. of Mathematics, Free University of Brussels, Belgium, 1974.

Lowen, M. R. Topologie generale. *C. R. Acad. Sci. Paris* (Serie A) **278**, 925–8, 1974.

Lowen, M. R. *Fuzzy Topological Spaces and Fuzzy Compactness*. Location and date unknown..

Lowen, P. Convergence Flous. *C. R. Acad. Sci. Paris*, 1974.

Lowen, R. Topologie flous. *C. R. Acad. Sci. Paris* 925–8.

Lukasiewicz, J. O logike trojwartosciowej. *Ruch. Fil. Varsovie* **169**, 1920.

Lukasiewicz, J., and Tarski, A., Untersuchungen uber den Aussagenkaikul. *C. R. Soc. Lett. Varsovie* **23**, 30–50, 1930.

McNaughton, R. A theorem about infinite-valued sentential logic. *J. Symbol. Logic* 1–13, 1951..

Maarschalk, C. G. D. Exact and fuzzy concepts superimposed to the GST: a meta theory. *3rd Int. Congr. of Cybernetics and Genese Systems, Bucharest, Roumania*, 1975.

MacVicar-Whelan, P. J. *Fuzzy Sets, the Concept of Height, and the Hedge "Very"*. Tech. Memo 1, Physics Dept., Grand Valley State Colleges, Allendale, Mich., 1974.

Machine, K. F. Vague predicates. *Am. Phil. Q.* 225–33, 1972.

Malvache, N. *Analyse et Identification des Systèmes Visuel et Manuel en Vision Frontale et Peripherique chez l'Homme*. These Doctorat en Sciences, Lille, France, 1973.

Malvache, N., and Illayes, D. *Representation et Minimisation de Fonctions Flous.* Doc. Centre Université de Valenciennes, France, 1974.

Malvache, N., Milbred, G., and Vidal, P. *Perception Visuelle: Champ de Vision Laterale, Modèle de la Fonction du Regard.* Rapp. Synthese, Contrat DRME 71-251, Paris, France, 1973.

Malvache, N., and Vidal, P. *Application des Systèmes Flous á la Modelisation des Phenomenes de Prise de Decision et d'Apprehension des Informations Visuelles chez l'Homme.* Rep. No. ATP-CNRS, 1K05, Paris, France, 1974..

Malvache, N., and Vidal, P. *Representation et Minimisation de Fonctions Flous.* Doc. Centre Université de Valenciennes, France, 1974.

Mamdani, E. H. Applications of fuzzy algorithms for control of simple dynamic plant. *Proc. IEE* 1585-8, 1974.

Mamdani, E. H. Application of fuzzy logic to approximate reasoning using linguistic synthesis, *Proc. 6th Int. Symp. on Multiple-Valued Logic, Logan, Utah, 1976.*

Mamdani, E. H. Application of fuzzy algorithms for the control of a dynamic plant. *Proc. IEE* in the press.

Mamdani, E. H. *IFAC Rep. on the 2nd Round Table Disc. on Fuzzy Automata and Decision Processes: Workshop on Discrete Systems and Fuzzy Reasoning.* Dept. of Electrical Engineering, University of Essex, Colchester, U.K., 1976.

Mamdani, E. H., and Assilian, S. An experiment in linguistic synthesis with a fuzzy logic controller. *Int. J. Man–Machine Stud.* 1–13, 1975.

Marinos, P. N. *Fuzzy Logic.* Bell Telephone Lab. Fuzzy Logic Case 36279-52, 39199-14, 1966.

Marinos, P. Fuzzy logic and its application to switching systems. *Trans. Computers* **C-18,** 343–8, 1969.

Maurer, W. D. *Input–Output Correctness and Fuzzy Correctness.* George Washington University, 1974..

Marks, P. *A Control System for Fuzzy Logic.* Thesis, Queen Mary College, London, 1975.

Meseguer, J., and Sols, I. Automata in semimodule categories. *Proc. 1st Int. Symp. on Category Theory Applied to Computation and Control,* 196–202, 1974.

Meseguer, J., and Sols, I. *Fuzzy Semantics in Higher Order Logic and Universal Algebra.* University of Zarogoza, Spain, 1975.

Meseguer, J., Sols, I. *Topology in Complete Lattices and Continuous Fuzzy Relations.* University of Zarogoza, Spain, 1975.

Michalski, R. S. Variable valued logic and its application to pattern recognition and machine learning. In *Multiple Valued Logic and Computer Science.* Amsterdam: North-Holland, 1975.

Mizumoto, M., Tanaka, K., and Toyoda, J. Fuzzy algebra. *Math. Sci. Res. Rec.* Kyoto University, 1971.

Mizumoto, M. *Fuzzy Automata and Fuzzy Grammars.* Thesis, Faculty of Engineering Science, Osaka University, Japan, 1971.

Mizumoto, M. Fuzzy set theory. *11th Prof. Group Meeting on Control Theory of SICE,* 1971.

Mizumoto, M., Toyoda, J., and Tanaka, K. Some considerations on fuzzy automata. *Trans. Electron. Commun. Engng. Japan* **52-C**, 93–9, 1969.

Mizumoto, M., Toyoda, J., and Tanaka, K. Some Considerations on fuzzy automata. *J. Computer Syst. Sci.* 409–22, 1969.

Mizumoto, M., Toyoda, J., and Tanaka, K. Fuzzy languages. *Trans. Electron. Commun. Engng. Japan* **53-C**, 333–40 (in Japanese). 1970.

Mizumoto, M., Toyoda, J., and Tanaka, K. L-fuzzy logic. *Research on Many-Valued Logic and Its Applications,* Kyoto University, Japan, 1972.

Mizumoto, M., Toyoda, J., and Tanaka, K. Formal grammars with weights. *Trans. IECE* **55-D**, 292–3, 1972.

Mizumoto, M., Toyoda, J., and Tanaka, K. General formulation of formal grammars. *Inform. Sci.* 87–100, 1972.

Mizumoto, M., Toyoda, J., and Tanaka, K. N-fold fuzzy grammars. *Inform. Sci.* 25–43, 1973.

Mizumoto, M., Toyoda, J., and Tanaka, K. Examples of formal grammars with weights. *Inform. Process. Lett.* 74–8, 1973.

Moisil, G. C. Recherches sur les logiques non-Chrysippiennes. *Annl. Sci. Univ. Jassy* 431–6, 1940.

Moisil, G. C. Logique modale. *Disq. Math. Phys.* **2**, 3–98, 1942.

Moisil, G. C. Sur les logiques de Lukasiewicz à un nombre fini de valeurs. *Rev. Roumaine Math. Pures Appl.* 905, 1964.

Moisil, G. C. *Incercari Vechi si noi de Logica Neclasica.* Stiintifica, Bucharest, 1965.

Moisil, G. C. Lukasiewiczian algebras. Centre de Calcul de l'Université de Bucharest, Roumania, 1968.

Moisil, G. C. Role of computers in the evolution of science. *Proc. Int. Conf. on Science and Society, Belgrade,* 134–6, 1971.

Moisil, G. C. *Essais sur les Logiques non Chrysippiennes.* Roumanian Academy of Sciences, Bucharest, Roumania, 1972.

Moisil, G. C. *Many-Values Logics and Hazard Phenomena in Switching Circuits.* University of Waterloo, Canada, 1973.

Mukaidono, M. *On Some Properties of Fuzzy Logic.* Tech. Rep. on Automation of IECE, 1972.

Mukaidono, M. On the B-ternary logical function—a ternary logic with consideration of ambiguity. *Trans. IECE* **55-D,** 355–62, 1972.

Mukaidono, M. *On Some Properties of Fuzzy Logic.* Tech. Rep. on Automation of IECE.

Nadiu, G. S. Sur la logique de Heytung. *Logique, Automatique, Informatique,* 42–70. Academie Republique Socialiste de Roumanie, 1971.

Nakata, H., Mizumoto, M., Toyoda, J., and Tanaka, K. Some characteristics of N-fold fuzzy CF grammars. *Trans. IECE Japan* **55-D,** 287–8, 1972.

Nahmias, S. *Discrete Fuzzy Random Variables.* University of Pittsburgh, Pittsburgh, Penn.

Narzaroff, G. J. Fuzzy topological polysystems, *J. Math. Analysis Applic.* 478–85, 1973.

Nasu, M., and Honda, N. Fuzzy events realized by finite probabilistic automata. *Inform. Control* 248–303, 1968.

Nasu, M., and Honda, N. Mapping induced by PGSM-mappings and some recursively unsolvable problems of finite probabilistic automata. *Inform. Control* 250–73, 1969.

Neff, T. P., and Kandel, A. *A New Approach to the Minimization of Fuzzy Switching Functions.* Rep. No. CSR 134, New Mexico Institute of Mining and Technology, Socorro, N.M., 1974.

Neff, T. P., and Kandel, A. Simplification of fuzzy switching functions. *Int. J. Computer Inform. Sci.* **6,** (1), 1977..

Negoita, C. V. On the application of the fuzzy sets separation theorem for automatic classification in information retrieval systems. *Inform. Sci.* 279–86, 1973.

Negoita, C. V., and Ralescu, D. A. *Fuzzy Sets and their Applications*. Technical Press, Bucharest, Roumania, 1974.

Negoita, C. V., and Ralescu, D. A. Fuzzy systems and artificial intelligence. *Cybernetics* 173–8, 1974.

Negoita, C. V., and Ralescu, D. A. Inexactness in Dynamic Systems. *Econ. Computation Econ. Cybernet. Stud. Res.* 69–81, 1974.

Negoita, C. V., and Ralescu, D. A. Relations on monoids and minimal realization theory for dynamic systems: applications for fuzzy systems. *Proc. 3rd Int. Congr. of Cybernetics and Systems, Bucharest, Roumania,* 1975.

Negoita, C. V., and Ralescu, D. A. Some results in fuzzy systems theory. *Proc. 3rd Int. Congr. on Cybernetics and General Systems, Bucharest, Roumania,* 1975.

Negoita, C. V., and Ralescu, D. A. Comment on "A comment on an algorithm that generates fuzzy prime implicants" by Lee and Chang, *Inst. Management Inform. Syst.,* 1976.

Netto, A. B. Fuzzy classes. *Not. Am. Math. Soc.* **68T-H28,** 945, 1968.

Noguchi, Y. A pattern clustering method on the basis of association schemes. *Bull. Electrotech. Lab.* 753–67, 1972.

Okada, N., and Tamachi, T. Automated editing of fuzzy line drawings for picture description. *Trans. IECE Japan* **57-A,** 216–23, 1974.

Okuda, T., Tanaka, H., and Asai, K. Decision-making and information in fuzzy events. *Bull. Univ. Osaka Prefect.* Ser. A, 1974.

Osis, J. J. Fault detection in complex systems using theory of fuzzy sets. *Kibernet. Diagnost.* 13–8, 1968.

Otsuki, S. A model for learning and recognizing machine. *Inform. Process.* 664–71, 1970.

Pask, G. Comments on fuzzy relations in representing cognitive structures. *Workshop on Discrete Systems and Fuzzy Reasoning.* Dept. of Electrical Engineering, Queen Mary College, London, 1976.

Paz, A. Fuzzy star functions, probabilistic automata and their approximation by nonprobabilistic automata. *J. Computer Syst. Sci.* 371–90, 1967.

Peschel, M. Some remarks to "Fuzzy systems" as a complement to the topic paper from L. A. Zadeh, Berlin, 1975.

Pinkava, G. Aspects of multivalued logics. *Workshop on Discrete Systems and Fuzzy Reasoning.* Dept. of Electrical Engineering, Queen Mary College, London, 1976.

Post, E. L. Introduction to a general theory of elementary propositions. *Am. J. Math.* 163–85, 1921.

Poston, T. *Fuzzy Geometry.* Thesis, University of Warwick, U.K., 1971.

Pratapa Reddy, V. C. V., and Ameling, W. Electronic implementation of fuzzy logic, to be published.

Preparata, F. P., and Yeh, R. T. *A Theory of Continuously Valued Logic.* Tech. Rep. No. 89, University of Texas, Austin, Texas, 1970.

Preparata, F. P., and Yeh, R. T. On a theory of continuously valued logic. *Conf. Rec. 1971 Symp. on the Theory and Applications of Multiple-Valued Logic Design,* 124–32, 1971.

Preparata, F. P., and Yeh, R. T. Continuously valued logic. *J. Computer Syst. Sci.* 397–418, 1972.

Prugovecki, E. A postulational framework for theories of simultaneous measurement of several observables. *Found. Phys.* **3,** 1973.

Prugovecki, E. Fuzzy sets in the theory of measurement of incompatible observables. *Found. Phys.* **4,** 9–18, 1974.

Prugovecki, E. *Measurement in Quantum Mechanics as a Stochastic Process on Spaces of Fuzzy Events,* Dept. of Mathematics, University of Toronto, Canada, 1974.

Pun, L. Experience in the use of fuzzy formalism in problems with various degrees of subjectivity. *Special Interest Discussion on Fuzzy Automata and Decision Processes, 6th IFAC World Congr., Boston, Mass.,* 1975.

Pun, L., and Doumiengts, G. Fuzzy decision process for solving production management problems in the industry of mechanical construction. *3rd Int. Conf. on Production Research, Amherst, Mass.*

Ragade, R. K. A multiattribute perception and classification of visual similarities. *Syst. Res. Planning Pap.* No. S-001-73, Bell-Northern Research, Ottawa, Canada, 1973.

Ragade, R. K. *On Some Aspects of Fuzziness in Communication. 1. Fuzzy Entropies.* Rep. No. W-002-73, Bell-Northern Research, Ottawa, Canada, 1973.

Ragade, R. K. *On Some Aspects of Fuzziness in Communication. 3. Fuzzy Concept Communication.* Bell-Northern Research, Ottawa, Canada, 1973.

Ragade, R. K. *A Note on Fuzzy Information Processing.* Bell-Northern Research, Ottawa, Canada, 1974.

Ragade, R. K. *Profile Transformation in Groups and Consensus Formation by Fuzzy Sets. 3. Fuzzy Concept Communication.* University of Waterloo, Waterloo, Ont., Canada, 1975.

Rajasethupathy, K. S., and Lakshmivarahan, S. *Connectedness in Fuzzy Topology.* Dept. of Mathematics, Vivekanamdha College, Madras, India, 1974.

Rajeck, R. K. *Benefit Cost Analysis under Imprecise Conditions.* University of Waterloo, Waterloo, Ont., Canada, 1975.

Reisinger, L. On fuzzy thesauri, Compstat 1974. *Proc. Conf. on Computational Statistics, Vienna,* 119–27, 1974.

Rickman, S. M., and Kandel, A. Column table approach for the minimization of fuzzy functions. Rep. No. CSR 137, Dept. of Computer Science, New Mexico Institute of Mining and Technology, Socorro, N.M., 1975.

Rieger, B. B. *On a Tolerance Topology Model of Natural Language Meaning.* Tech. Rep. Germanic Institute, Technische Hochschule, Aachen, Federal Republic of Germany, 1975.

Roberts, F. S. Tolerance Geometry. *Notre Dame J. Formal Logic* 68–76, 1973.

Rodder, W. *On "AND" and "OR" Connectives in Fuzzy Set Theory.* Rep. No. Euro I 75/07, Lehrstuhl für Unternehmensforschung Rwth, Aachen, Federal Republic of Germany, 1975.

Rosenfeld, A. Fuzzy Groups. *J. Math. Analysis Applic. 512*–7, 1971.

Rosenfeld, A. Fuzzy graphs. In *Fuzzy Sets and their Applications to Cognitive and Decision Processes* (Ed. L. A. Zadeh *et al.*), 77–96, Academic Press, New York, 1975.

Ruspini, E. H. A new approach to clustering. *Inform. Control* 22–32, 1969.

Ruspini, E. H. Numerical methods for fuzzy clustering. *Inform. Sci.* 319–50, 1970.

Ruspini, E. H. Optimization in sample descriptions: data reduction and pattern recognition using fuzzy clustering (abstr.). *IEEE Trans. Syst. Man Cybernet.* **SMC-2,** 541, 1972.

Ruspini, E. H. New experimental results in fuzzy clustering. *Inform. Sci.* 273–84, 1973.

Ruspini, E. H. *A Fast Method for Probabilistic and Fuzzy Cluster Analysis using Association Measures.* Space Biology Lab., Brain Research Institute, University of California, Los Angeles, Calif., 1973.

Russell, B. Vagueness, *Aust. J. Phil.* 84–92, 1923.

Rutherford, D. A., and Bloore, G. C. *The Implementation of Fuzzy Algorithms for Control.* Control Systems Centre, University of Manchester Institute of Science and Technology, U.K., 1975.

Rutherford, D. A. Implementation and evaluation of a fuzzy control algorithm in a sinter plant. *Workshop on Discrete Systems and Fuzzy Reasoning.* Dept. of Electrical Engineering, Queen Mary College, London, 1976.

Sanchez, E. *Equations de Relations Flous.* Ph.D. Thesis, Faculté de Medecin de Marseille, Marseille, France, 1974.

Sanchez, E. Resolution of composite fuzzy relation equations. *Inform. Control* **30**, 38–48.

Sanford, D. H. Infinity and vagueness, *Phil. Rev.* 520–35, 1975.

Sanford, D. H. Borderline logic. *Am. Phil. Q.* 29–39, 1975.

Santos, E. S. Maximin automata, *Inform. and Control* 363–77, 1968.

Santos, E. S. Maximin sequential-like machines and chains. *Math. Syst. Theor.* **3**, 300–9, 1969.

Santos, E. S. Fuzzy algorithms. *Inform. Control* 326–39, 1970.

Santos, E. S. On reduction of maximin machines. *J. Math. Analysis Applic.*, 1972.

Santos, E. S. Fuzzy algorithms. *Inform. Control* 1–11, 1974.

Santos, E. S. Fuzzy automata and languages. *US–Japan Seminar on Fuzzy Sets and their Applications, Berkeley, Calif., 1974.*

Santos, E. S. Realization of fuzzy languages by probabilistic, max-product and maxim automata. *Inform. Sci.* 39–53, 1975.

Santos, E. S. Fuzzy programs. *6th IFAC World Congr., Boston, Mass.,* 1975.

Santos, E. S. Max-product grammars and languages. *Inform. Sci.* 1–23, 1975.

Santos, E. S. Fuzzy sequential functions. *J. Cybernet.* in the press.

Santos, E. S., and Wee, N. G. General formulation of sequential machines, *Inform. Control* 5–10, 1970.

Saridis, G. N. Fuzzy decision making in prosthetic devices and other applications. *Special Interest Discussion on Fuzzy Automata and Decision Processes, 6th IFAC World Congr., Boston, Mass.,* 1975.

Saridis, G. N. *Fuzzy Notions in Nonlinear System Classification.* Purdue University, West Lafayette, Ind., 1975.

Sasama, H. *Fuzzy Set Model for Train Composition in Marshalling Yard, Rep. 1. Working Group on Fuzzy Systems,* 49–54, Tokyo, Japan, 1975.

Schimura, M. Fuzzy sets concept in rank-ordering objects. *J. Math. Analysis Applic.* 717–33, 1973.

Schutzenberger, M. P. On a theory of P. Jungen. *Proc. Am. Math. Soc.* 885–90, 1962.

Schwede, G. N-variable fuzzy maps with application to disjunctive decomposition of fuzzy switching functions. *Proc. 6th Int. Symp. on Multiple-Valued Logic, Logan, Utah, 1976.*

Serfati, M. *Algèbres de Boole avec une Introduction à la Theorie des Graphes Orientes et aux Sous-Ensembles Flous.* C. D. U., Paris, 1974.

Serizawa, M. A search technique of control rod pattern of smoothing core power distributions by fuzzy automation. *J. Nucl. Sci. Technol.,* 1973.

Shimura, M. Fuzzy sets concept in rank-ordering objects. *J. Math. Analysis Applic.* 717–33, 1973.

Shimura, M. Application of fuzzy functions to pattern classification. *Trans. IECE Japan,* **55-D,** 218–25.

Shimura, M. Fuzzy sets in rank-ordering objects, *J. Math. Analysis. Applic.* 717–33, 1973.

Shimura, M. An approach to pattern recognition and associative memories using fuzzy logic. *US–Japan Seminar of Fuzzy Sets and their Applications, Berkeley, Calif.,* 1974.

Shimura, M. An approach to pattern recognition and associative memories using fuzzy logic. In *Fuzzy Sets and their Applications to Cognitive and Decision Processes* (Ed. L. A. Zadeh *et al.*), 449–76. Academic Press, New York.

Sinha, N. K., and Wright, J. D. Application of fuzzy control to a heat exchanger. *Special Interest Discussion on Fuzzy Automata and Decision Processes, 6th IFAC World Congr., Boston, Mass.,* 1975.

Siy, P. *Fuzzy Logic and Handwritten Numeral Character Recognition.* Thesis, Dept. of Electrical Engineering, University of Akron, Akron, Ohio, 1973.

Siy, P., and Chen, C. S. *Fuzzy Logic Approach to Handwritten Character Recognition Problem.* Dept. of Electrical Engineering, University of Akron, Akron, Ohio, 1971.

Siy, P., and Chen, C. S. Minimization of fuzzy functions. *IEEE Trans. Computers,* 100–2, 1972.

Siy, P., and Chen, C. S. Fuzzy logic for handwritten numeral character recognition. *IEEE Trans. Syst. Man Cybernet.* 570–5, 1974.

Skala, H. J. *On the Problem of Imprecision.* Dordrecht, Netherlands, 1974.

Smith, R. E. *Measure Theory on Fuzzy Sets.* Thesis, Dept. of Mathematics, University of Saskatchewan, Saskatoon, Canada, 1970.

Sols, I. *Topology in Complete Lattices and Continuous Fuzzy Relations.* Zaragoza University, Zaragoza, Spain, 1975.

Stoica, M., and Scarlat, E. Fuzzy concepts in the control of production systems. *Proc. 3rd Int. Congr. of Cybernetics and Systems, Bucharest,* 1975.

Stone, M. H. Topological representation of distributive lattices and Brouwerian logics. *Čas. Math. Phys.* 1–25, 1975.

Sugeno, M. On fuzzy nondeterministic problems, *Annual Conf. Rec. SICE,* 1971.

Sugeno, M. Fuzzy measures and fuzzy integrals. *Trans. SICE,* 218–26, 1972.

Sugeno, M. Constructing fuzzy measure and grading similarity of patterns by fuzzy integrals. *Trans. SICE* 359–67, 1973.

Sugeno, M. Evaluation of similarity of patterns by fuzzy integrals. *Annual Conf. Rec. SICE.*

Sugeno, M. *Theory of Fuzzy Integrals and its Applications.* Thesis, Tokyo Institute of Technology, Tokyo, Japan, 1974.

Sugeno, M., and Terano, T. An approach to the identification of human characteristics by applying fuzzy integrals. *Proc. 3rd IFAC Symp. on Identification and System Parameter Estimation,* 1973.

Sugeno, M., and Terano, T. Analytical representation of fuzzy systems. *Special Interest Discussion on Fuzzy Automata and Decision Processes, 6th IFAC World Congr., Boston, Mass.,* 1975.

Sugeno, M., Tsukamoto, Y., and Terano, T. Subjective evaluation of fuzzy objects. *Reprint, IFAC Symp. on Stochastic Control,* 1974.

Sugeno, S. Fuzzy systems and pattern recognition. *Workshop on Discrete Systems and Fuzzy Reasoning.* Dept. of Electrical Engineering, Queen Mary College, London, 1976.

Tahani, V. *Fuzzy Sets in Information Retrieval.* Thesis, Dept. of Electrical Engineering and Computer Science, University of California, Berkeley, Calif., 1971.

Tamura, S. Fuzzy pattern classification. *Proc. Symp. on Fuzziness in Systems and its Processing (SICE),* 1971.

Tamura, S., Higuchi, S., and Tanaka, K. Pattern lassification based on fuzzy relations. *Trans. IECE Japan* **53-C,** 937–44, 1970.

Tamura, S., Higuchi, S., and Tanaka, K. Pattern classification based on fuzzy relations. *IEEE Trans. Syst. Man Cybernet.* 98, 1973.

Tamura, S., and Tanaka, K. Learning of fuzzy formal language. *IEEE Trans. Syst. Man Cybernet.* 98–102, 1973.

Tanaka, H., and Kaneku, S. A fuzzy decoding procedure on error correcting codes. *Tech. Rep. Inform. IECE.* 1973.

Tanaka, H., Okuda, T., and Asai, K. On the fuzzy mathematical programming. *Annual Conf. Rec. SICE,* 1972.

Tanaka, H., Okuda, T., and Asai, K. Fuzzy mathematical programming. *Trans. SICE* **9,** 109–15, 1973.

Tanaka, H., Okuda, T., and Asai, K. Decision-making and its goal in a fuzzy environment. *US–Japan Seminar on Fuzzy Sets and their Applications, Berkeley, Calif.,* 1974.

Tanaka, H., Okuda, T., and Asai, K. *On Fuzzy-Mathematical Programming.* University of Osaka Prefecture, Japan.

Tanaka, H., Okuda, T., and Asai, K. *A Formulation of Fuzzy Decision Problems and its Application to an Investment Problem.* Dept. of Industrial Engineering, University of Osaka, Prefecture Sakai, Osaka, Japan, 1975.

Tanaka, K. Analogy and fuzzy logic. *Math. Sci.,* 1972.

Tanaka, K., and Mizumoto, M. Fuzzy programs and their execution. In *Fuzzy Sets and their Applications to Cognitive and Decision Processes* (Ed. L. A. Zadeh *et al.*), 41–76. Academic Press, New York, 1975.

Tanaka, K., Toyoda, J., and Mizumoto, H. Fuzzy automata theory and its application to automatic controls. *J. JAACE* 541–50, 1970.

Tanaka, K., Toyoda, J., Mizumoto, M., and Tsuji, H. Fuzzy automata theory and its application to automatic controls. *J. JAACE* 541–50, 1970.

Tazaki, E. Heuristic synthesis in a class of systems by using fuzzy automata. *Rep. 1, Working Group on Fuzzy Systems, Tokyo, Japan,* 61–6, 1975.

Terano, I. *Fuzziness of Systems* 21–5. Nikka-Giren Engineers, 1972.

Terano, T. Fuzziness and its concept. *Proc. Symp. on Fuzziness in Systems and its Processing, SICE.*

Terano, T., and Sugeno, M. Conditional fuzzy measures and their applications. In *Fuzzy Sets and their Applications to Cognitive and Decision Processes* (Ed. L. A. Zadeh *et al.*), 151–70. Academic Press, New York, 1975.

Terano, T., and Sugeno, M. Macroscopic optimization by using conditional fuzzy measures. *Rep. 1, Working Group on Fuzzy Systems, Tokyo, Japan*, 67–72, 1975.

Terano, T., and Sugeno, M. Macroscopic optimization by using conditional measures. *6th IFAC World Congr., Boston, Mass.*, 1975.

Thomason, M. G. Fuzzy syntax-directed translations. *J. Cybernet.* 87–94, 1974.

Thomason, M. G. *Finite Fuzzy Automata, Regular Fuzzy Languages, and Pattern Recognition.* Dept. of Electrical Engineering, Duke University, Durham, N. C., 1974.

Thomason, M. G. The effect of logic operations and fuzzy logic distributions. *IEEE Trans. Syst. Man Cybernet.* **SMC-5**, 309–10, 1974.

Thomason, M. G., and Marinos, P. N. Fuzzy logic relations and their utility in role theory. *Proc. Conf. on Systems, Man and Cybernetics, Washington, D.C.*, 1972.

Thomason, M. G., and Marinos, P. N. Deterministic acceptors of regular fuzzy languages. *IEEE Trans. Syst. Man Cybernet.* **SMC-4**, 228–30, 1974.

Tsichritzis, D. *Fuzzy Properties and Almost Solvable Problems.* Tech. Rep. No. 70, Dept. of Electrical Engineering, Princeton University, Princeton, N.J., 1968.

Tsichritzis, D. *Measures on Countable Sets,* Tech. Rep. No. 8, Dept. of Computer Science, University of Toronto, Canada, 1969.

Tsuji, H., Mizumoto, M., Toyoda, J., and Tanaka, K. Interaction between random environments and fuzzy automata with variable structures. *Trans. IECE Japan* **55-D**, 143–4, 1972.

Tsuji, H., Mizumoto, M., Toyoda, J., and Tanaka, K. Linear fuzzy automation. *Trans. IECE Japan,* **56-A**, 256–7, 1973.

Tsukamoto, Y. Identification of preference measure by means of fuzzy integrals. *JORS* preprint, 1972 (in Japanese).

Tsukamoto, Y. A subjective evaluation on attractivity of sightseeing zones. *Rep. 1, Working Group on Fuzzy Systems, Tokyo, Japan,* 1975.

Tsukamoto, Y., and Lida, H. Evaluation models of fuzzy systems. *Annual Conf. Rec. SICE,* 1973.

Uhr, L. Toward integrated cognitive systems, which must make fuzzy decisions about fuzzy problems. In *Fuzzy Sets and their Applications to Cognitive and Decision Processes* (Ed. L. A. Zadeh *et al.*), 353–94. Academic Press, New York, 1975.

Van Frassen, B. C. Comments: Lakoff's fuzzy propositional logic. In *Contemporary Research in Philosophical Logic and Linguistic Semantics*. (Eds. D. Hockney, W. Harper, and B. Freed). Reidel, Holland, 1975.

Van Velthoven, G. *Onderzoek naar Toepasbeerheid van de Theorie der Vage Verzamelingen op het Parametrisch Onderzoek Inzake Criminaliteit,* 1974.

Van Velthoven, G. *Application of Fuzzy Sets Theory to Criminal Investigation.* Euro I, Brussels, 1975.

Van Velthoven, G. Fuzzy models in personal management, *3rd Int. Congr. of Cybernetics and Systems, Bucharest, Roumania,* 25, 1975.

Van Velthoven, G. *Quelques Applications de la Taxonomie Flou.* Université des Sciences et Techniques de Lille, France.

Verma, P. P. Vagueness and the principle of the excluded middle. *Mind,* 66–77, 1970.

Vincke, P. *Une Application de la Theorie des Graphes Flous.* Université Libre de Bruxelles, 1973.

Vincke, P. La Theorie des Ensembles Flous, Memorie. Faculté de Science, Université Libre de Bruxelles, 1973.

Warren, R. H. *Closure Operator and Boundary Operator for Fuzzy Topological Spaces.* Applied Mathematics Research Lab., Wright-Patterson AFB, Ohio.

Warren, R. H. *Neighborhoods, Bases and Continuity in Fuzzy Topological Spaces.* Applied Mathematics Research Lab., Wright-Patterson AFB, Ohio.

Warren, R. H. *Optimality in Fuzzy Topological Polysystems.* Applied Mathematics Research Lab., Wright-Patterson AFB, Ohio, 1974.

Watanabe, S. Modified concepts of logic probability and information based on generalized continuous characteristic function. *Inform. Control* 1–21, 1969.

Wechler, W. Analyse und Synthes zeitvariabler R-fuzzy automaten. *Akad. Dt. Wiss. DDR,* 1974.

Wechler, W. *R-Fuzzy Grammars.* Technical University of Dresden, German Democratic Republic, 1974.

Wechler, W. R-Fuzzy automata with a time-variant structure. In *Lecture Notes in Computer Science,* Vol. 28, 73–6. Springer-Verlag, Berlin, 1975.

Wechler, W. *The Concept of Fuzziness in the Theory of Automata.* Technical University of Dresden, German Democratic Republic, 1975.

Wechler, W., and Dimitrov, V. R-Fuzzy Automata, *Proc. IFIP Congr.,* 657–60, 1974.

Wechsler, H. *Applications of Fuzzy Logic to Technical Diagnosis.* Information and Computer Sciences Dept., University of California, 1975.

Wee, W. G., and Fu, K. S. A formulation of fuzzy automata and its application as a model of learning systems, *IEEE Trans. Syst. Sci. Cybernet.* **SSC-5,** 215–23, 1969.

Wee, W. G., and Fu, K. S. *On a Generalization of Adaptive Algorithms and Applications of the Fuzzy Set Concept to Pattern Classification.* Tech. Rep. No. 67-7, Dept. of Electrical Engineering, Purdue University, West Lafayette, Ind., 1969.

Wee, W. G., and Fu, K. S. A formulation of fuzzy automata and its application as a model of learning systems. *IEEE Trans. Syst. Sci. Cybernet.* **SSC-5,** 215–23, 1975.

Weiss, M. D. Fixed points, separation and induced topologies for fuzzy sets. *J. Math. Analysis Applic.* 142–50, 1975.

Wenstop, F. *Applications of Linguistic Variables in the Analysis of Organizations.* Thesis proposal, University of California, Berkeley, Calif., 1974.

Wong, C. K. Covering properties of fuzzy topological spaces. *J. Math. Analysis Applic.* 697–704, 1973.

Wong, C. K. Fuzzy points and local properties of fuzzy topology. *J. Math. Analysis Applic.* 316–28, 1974.

Wong, C. K. Fuzzy topology: product and quotient theorems. *J. Math. Analysis Applic.* 512–21, 1974.

Wong, C. K. *Categories of Fuzzy Sets and Fuzzy Topological Spaces.* Rep. No. RC 5138, IBM Thomas B. Watson Research Center, Yorktown Heights, N.Y., 1974.

Wong, C. K. Fuzzy topology. In *Fuzzy Sets and their Applications to Cognitive and Decision Processes* (Ed. L. A. Zadeh *et al.*), 171–90. Academic Press, New York, 1975.

Wong, G. A., and Shen, D. C. On the learning behaviour of fuzzy automata. *Proc. Int. Congr. on Cybernetics and Systems, Oxford,* 1972.

Woodhead, R. G. *On the Theory of Fuzzy Sets to Resolve Ill-Structured Marine Decision Problems,* Dept. of Naval Architecture and Shipbuilding, University of Newcastle-upon-Tyne, U.K.

Yeh, R. T. *Toward an Algebraic Theory of Fuzzy Relational Systems.* Tech. Rep. No. TR-25, Dept. of Computer Science, University of Texas, Austin, Texas, 1973.

Yeh, R. T., and Bang, S. Y. Fuzzy relations, fuzzy graphs, and their applications to clustering analysis. In *Fuzzy Sets and their Applications to Cognitive and Decision Processes* (Ed. L. A. Zadeh et al.), 125–49. Academic Press, New York, 1975.

Yoeli, M. A note on a generalization of Boolean matrix theory. *Am. Math. Mon.* 552–57, 1961.

Zadeh, L. A. Fuzzy sets. *Inform. Control* 338–53, 1965.

Zadeh, L. A. Fuzzy sets and systems. *Proc. Symp. on System Theory,* 29–37. Polytechnic Press of the Institute of Brooklyn, New York, 1965.

Zadeh, L. A. *Shadows of Fuzzy Sets.* Memo ERL-M336, University of California, Berkeley, 1966.

Zadeh, L. A. Fuzzy algorithms. *Inform. Control* 94–102, 1968.

Zadeh, L. A. Probability measures of fuzzy events. *J. Math. Analysis Applic.* 421–7, 1968.

Zadeh, L. A. Machine intelligence *versus* human intelligence. *Proc. Conf. on Science and Society, Herceg-Novi, Yugoslavia,* 127–34, 1969.

Zadeh, L. A. *Toward Fuzziness in Computer Systems, Fuzzy Algorithms and Languages.* University of California, Berkeley, Calif., 1969.

Zadeh, L. A. Biological application of the theory of fuzzy sets and systems. In *Biocybernetics of the Central Nervous System* (Ed. L. D. Proctor), 199–212. Little, Brown, Boston, 1970.

Zadeh, L. A. Fuzzy languages and their relation to human and machine intelligence. *Proc. Conf. on Man and Computer, Bordeaux, France,* 1970.

Zadeh, L. A. Toward fuzziness in computer systems—fuzzy algorithms and languages. In *Architecture and Design of Digital Computers,* (Ed. G. Boulaye), 9–18. Dunod, Paris, 1971.

Zadeh, L. A. Quantitative fuzzy semantics. *Inform. Sci.* 159–76, 1971.

Zadeh, L. A. Toward a theory of fuzzy systems. In *Aspects of Network and System Theory.* Holt, Rinehart, and Winston, New York, 1971.

Zadeh, L. A. Similarity relations and fuzzy orderings. *Inform. Sci.* 177–200, 1971.

Zadeh, L. A. *A System-Theoretic View of Behaviour Modification. Memo. ERL-M320,* University of California, Berkeley, Calif., 1972.

Zadeh, L. A. *On Fuzzy Algorithms,* Memo. ERL-M325, University of California, Berkeley, Calif., 1972.

Zadeh, L. A. A fuzzy-set-theoretic interpretation of linguistic hedges. *J. Cybernet.* 4–34, 1972.

Zadeh, L. A. A rationale for fuzzy control. *J. Dynamic Syst. Meas. Control* 3–4, 1972.

Zadeh, L. A. Linguistic cybernetics. *Proc. Int. Symp. on Systems Sciences and Cybernetics, Oxford,* 1972.

Zadeh, L. A. Mathematics—a call for reorientation. *Math. Sci.,* 1972.

Zadeh, L. A. Outline of a new approach to the analysis of complex systems and decision processes. *IEEE Trans. Syst. Man Cybernet.* **SMC-3,** 28–44, 1973.

Zadeh, L. A. On the analysis of large-scale systems. In *Systems Approaches and Environmental Problems, Int. Symp. Schlob Reisenburg.* Gunzburg, Ulm, 1973.

Zadeh, L. A. *The Concept of a Linguistic Variable and its Application to Approximate Reasoning.* Memo. ERL-M411, University of California, Berkeley, Calif., 1973 (also published in *Inform. Sci.,* 1975.)

Zadeh, L. A. *On the Analysis of Large-Scale Systems.* Memo. ERL-M418, University of California, Berkeley, Calif., 1974.

Zadeh, L. A. Numerical *versus* linguistic variables. *Newslett. Circuits Syst. Soc.* 3–4, 1974.

Zadeh, L. A. Fuzzy logic and its application to approximal reasoning. *Proc. IFIP Congr., Stockholm, Sweden,* 1974.

Zadeh, L. A. *A Fuzzy-Algorithmic Approach to the Definition of Complex or Unprecise Concepts.* Memo. ERL-M474, University of California, Berkeley, Calif., 1974.

Zadeh, L. A. *Fuzzy Logic and Approximate Reasoning.* Memo. ERL-M479, University of California, Berkeley, Calif., 1974.

Zadeh, L. A. Calculus of fuzzy restrictions. In *Fuzzy Sets and Their Applications to Cognitive and Decision Processes* (Ed. L. A. Zadeh *et al.*), 1–40. Academic Press, New York, 1975.

Zadeh, L. A. A relational model for approximate reasoning. *IEEE Int. Conf. on Cybernetics and Society, San Francisco, Calif.,* 1975.

Zadeh, L. A. Semantic influence from fuzzy premises. *6th Int. Symp. on Multiple Valued Logic, Logan, Utah,* 217–18, 1976.

Zadeh, L. A., Fu, K. S., Tanaka, K., and Shimura, M. *Fuzzy Sets and their Applications to Cognitive and Decision Processes.* Academic Press, New York, 1975.

Zimmermann, H. J. Optimization in fuzzy environments. *Int. TIMS and 46th ORSA Conf., San Juan, Puerto Rico,* 1974.

Zimmermann, H. J. *Optimale Entscheidungen bei Unscharfen Problembeschreibungen.* Lehrstuhl für Unternahmensforschung, RWTH Aachen, Federal Republic of Germany, 1975.

Zimmermann, H. J. Description and optimization of fuzzy systems. *Int. J. Gen. Syst.* **2,** 209–15, 1975.

Zimmermann, H. J. *The Potential of Fuzzy Decision-Making in the Private and Public Sector.* Doc. S Oak-75, Lidingø, Sweden, 1975.

Zimmermann, H. J. *Bibliography: Theory and Applications of Fuzzy Sets,* 75/16, Institüt für Wirtschaftawissenschaften, Aachen, Federal Republic of Germany, 1975.

Zimmermann, H. J., and Gehring, H. Fuzzy information profile for information selection. *4th Int. Congr., Paris, 1974.*

Index